SpringerBriefs in Materials

The SpringerBriefs Series in Materials presents highly relevant, concise monographs on a wide range of topics covering fundamental advances and new applications in the field. Areas of interest include topical information on innovative, structural and functional materials and composites as well as fundamental principles, physical properties, materials theory and design. SpringerBriefs present succinct summaries of cutting-edge research and practical applications across a wide spectrum of fields. Featuring compact volumes of 50 to 125 pages, the series covers a range of content from professional to academic. Typical topics might include:

- A timely report of state-of-the art analytical techniques
- A bridge between new research results, as published in journal articles, and a contextual literature review
- A snapshot of a hot or emerging topic
- An in-depth case study or clinical example
- A presentation of core concepts that students must understand in order to make independent contributions Briefs are characterized by fast, global electronic dissemination, standard publishing contracts, standardized manuscript preparation and formatting guidelines, and expedited production schedules.

More information about this series at http://www.springer.com/series/10111

Jiajun Gu • Di Zhang • Yongwen Tan

Metallic Butterfly Wing Scales

Superstructures with High Surface-Enhancement Properties for Optical Applications

Springer

Jiajun Gu
State Key Lab of Metal Matrix Composites
Shanghai Jiao Tong University
Shanghai
China

Yongwen Tan
State Key Lab of Metal Matrix Composites
Shanghai Jiao Tong University
Shanghai
China

Di Zhang
State Key Lab of Metal Matrix Composites
Shanghai Jiao Tong University
Shanghai
China

ISSN 2192-1091   ISSN 2192-1105 (electronic)
SpringerBriefs in Materials
ISBN 978-3-319-12534-3   ISBN 978-3-319-12535-0 (eBook)
DOI 10.1007/978-3-319-12535-0

Library of Congress Control Number: 2014956390

Springer Cham Heidelberg New York Dordrecht London

Printed on acid-free paper

Springer is part of Springer Science+Business Media (www.springer.com)

Preface

By a long-term evolution, biological species have developed numerous powerful structures and functions to survive the natural selection process. These natural designs always present inspirations for solving complex human problems. Early imitations of biological structures can be traced back to six centuries ago, when Leonardo da Vinci tried to fabricate a "flying machine" by studying the anatomy and flight of birds. However, although scientists have never stopped artificially imitating nature (*biomimetics*), many natural designs still surpass the most advanced human products in complexity and functionality.

Over a decade ago, however, a group of new materials, morphology genetic materials, was developed (Cook et al., Angew Chem Int Ed 2003, 42, p. 557). Using a great variety of chemical and physical methods, scientists directly replicated the bio-structures in functional components. These materials not only preserve the original biological morphologies, but also have a broad range of manually endowed functions.

As a part of these studies, this book mainly demonstrates how natural butterfly wing scales can be used to produce advanced SERS substrates. These scales feature 3D sub-micrometer structures that are superior to many human designs in terms of structural complexity, reproducibility, and cost. Moreover, there are globally over 174,500 species of butterflies and moths and each bears various types of scales. This fact presents much freedom for researchers to select appropriate structures according to different applications.

This book presents three approaches for replicating natural butterfly wing scales using a variety of metals required by state-of-the-art science and technology. Among these methods, a single versatile chemical route (Chap. 3) can be used to produce butterfly scales in seven different metals (Ag, Au, Cu, Co, Ni, Pd, and Pt). Significantly, the Au scales such as SERS substrates have ten times the analyte detection sensitivity and are one tenth the cost of their human-designed commercial counterparts (Klarite™). Since some species of butterflies and moths (e.g., silk moth) have been farmed for thousands of years and one wing usually bears over 100,000 scales, these materials are promising in SERS applications. In addition, preliminary mechanisms of these surface-enhancement phenomena are also included in this book, which have been applied in designing artificial arrays with high SERS performance (Jeon et al., Small 2014, 10, p. 1490).

The main contents of this book are based on the studies conducted in the past 5 years in the Morphology Genetic Lab established by Prof. Di Zhang at Shanghai Jiao Tong University. These replications of butterfly scales to metals were carried out after the first metal butterfly made via PVD in the Vukusic group at the University of Exeter (Garrett et al., J Biophoton 2009, 2, p. 157). We greatly appreciate Mr. Xiang Ding, Mrs. Jing Fang, Mrs. Guofen Song, and Mr. Wenshu Chen for editing the drafts. We would also like to thank the financial support from the Natural Science Foundation of China (No. 51202145, No. 51171110, No. 51271116, and No. 91333115), Shanghai Science and Technology Committee (No. 14JC1403300 and No. 14520710100), and Research Fund for the Doctoral Program of Higher Education of China (20120073120006 and 20120073130001). In addition, I appreciate the support provided by the Program for New Century Excellent Talents in University, Ministry of Education, China.

Finally, I would like to greatly thank Ms. Sara Kate Heukerott at Springer for her kind help and patience in handling this book. I also wish to thank my family for their consistent support in the past year, which helped me recover from the knee fracture.

26 Aug 2014   Jiajun Gu
                                                                       Shanghai Jiao Tong University

Contents

Chapter 1
Background

In the twentieth century, the understandings of materials' properties deepened to an electronic level. Present technologies have realized the control of the electron motion in media based on the advances in semiconductor physics. This in turn leads to a rapid development in electronics and information technologies (Moore's law). However, an information processor needs high speed and capacity while the capacity of an electronic device is limited. As Moore's law might be approaching its limits, scientists are seeking for alternative strategies to satisfy the rapidly increased application demands. One of the most promising solutions is to use photons as information carriers to replace electrons. Over the past decades, great efforts have been spent in seeking ways to effectively control light propagation within materials. These studies have opened up a new research field; nanophotonics.

Currently, photonic crystals and plasmonic devices are two of the most promising solutions for the manipulation of light at a submicrometer level. Photonic crystals (or photonic band gap materials) refer to a group of artificial dielectric structures with photonic band gaps [1]. Radiations with specific wavelength can propagate in these structures while others are prohibited. This is achieved by a periodic modulation in dielectric coefficients by materials and environments. Analogous to semiconductors with electronic band gaps, photonic crystals are critical to optical integrated circuits. Optical components like photonic crystal fibers, photonic crystal optical waveguides, photonic crystal couplers, and photonic crystal beam splitters have been developed so far [2].

However, typical photonic structures are of hundreds of nanometers in dimension. Though photonic devices have high speeds and broad bandwidths, their size is comparable to light wavelength and the energy loss will rapidly increase with the decrease in device dimension. Thus, it is difficult to interconnect photonic structures with conventional nanosized electronic devices. To solve this problem, plasmonic devices were recently developed [3–7]. Since the dielectric coefficients of *metals* are *negative*, metallic nanostructures can generate surface plasmon polaritons (SPPs) or localized surface plasmons (LSPs) under the excitation of electromagnetic waves [8]. These SPPs or LSPs within a subwavelength region can carry information instead of mere electrons or photons, thus bridging the structural gap

© Jiajun Gu, Di Zhang, and Yongwen Tan 2015
J. Gu et al., *Metallic Butterfly Wing Scales,* SpringerBriefs in Materials,
DOI 10.1007/978-3-319-12535-0_1

between traditional photonic structures and electronic devices. Plasmonic devices are promising in light convergence, light waveguide, and light tunneling, etc., which may greatly promote state-of-the-art "optical + electronic" integration technologies. Moreover, since the SPPs and LSPs can cause localized energy enhancement, they can be used to amplify optical signals on metal surface. In the following section, we will briefly review how these materials are used and how they can be fabricated at present.

1.1   Plasmonic Structures

When a light beam irradiates on a metal surface, it will interact with the free electrons of the metal under certain conditions to form a series of excited states (Fig. 1.1). These states are known as SPPs [8, 9]. In addition, once the light interacts with rough metal structures or nanoparticles (NPs), another type of bound state, i.e., LSPs, will be generated on the metal surfaces [9]. Both SPPs and LSPs are near-field effects. The strong interactions between the light and metals can be regulated by changing the metal species (especially Ag, Au, and Cu), periodicity of metal structures, size and shape of metal particles, and surrounding environments, etc. As mentioned above, they have a broad range of applications in trace-detections [10–15] and optics (e.g., in integrated optical circuits; [3, 16–18]).

In chemical and biological detections, metallic submicrometer structures can mainly be used in two ways. First, metallic submicrometer structures or metal NPs can be applied as surface-enhanced Raman scattering (SERS) substrates. Raman signal of analytes immobilized on these supports can be drastically amplified. SERS substrates are powerful tools for trace-detections and have wide applications in biomedical technology, environmental engineering, social security, chemical industry, etc. [19–24]. Some supports with specially textured surface can even achieve single-molecule detection. Second, metallic structures can be used in fluorescence detections. Metallic structures can drastically enhance the fluorescence signals of fluorophores absorbed on nanostructured metal surface, which is especially helpful

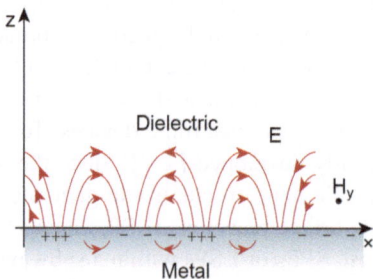

Fig. 1.1 Surface plasmon polaritons (SPPs). (Reproduced with permission [3] Copyright 2003, Nature Publishing Group)

to bioscientists [25–28]. Such a metal-enhanced fluorescence (MEF) phenomenon is also referred as surface-enhanced fluorescence (SEF). In general, both SERS and MEF can be optimized by tuning the localized surface plasmon resonance (LSPR) on the supporting material [29].

In integrated circuits, plasmonics may provide potential solutions for development of new photonic devices, broadband communication systems, tiny scale photonic circuits, optical sensors and detectors, etc. [3, 16–18].

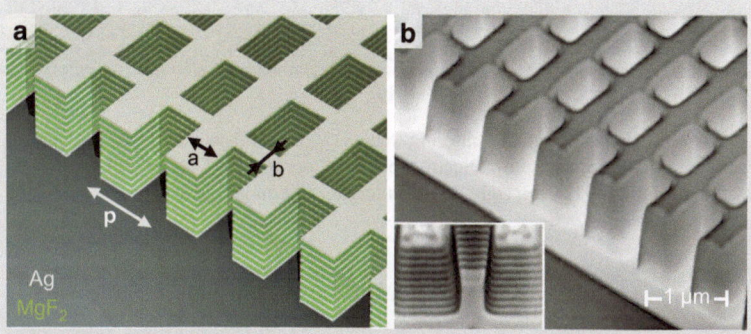

Box 1.1   A negative index metamaterial. Reproduced with permission [30], Copyright 2008, Nature Publishing Group.

It should be especially noted that based on the negative dielectric coefficients of metal, scientists have also developed a group of interesting materials known as negative index materials (NIMs) or metamaterials [31–33]. In 1999, J. B. Pendry theoretically proved that negative permittivity and negative permeability can be obtained from well-designed metal photonic structures or metal split ring resonators [31]. In 2001, D. R. Smith et al. first observed negative refractions in two-dimensional (2D) metal arrays within a microwave-band [32]. In 2008, X. Zhang et al. achieved negative refraction for three-dimensional (3D) metal arrays within a visible-light-band [30]. In the same year, V. M. Shalaev provided the principles for "invisible cloaks" made of metamaterials [34]. Since NIMs exhibit completely different optical properties from their traditional counterparts with positive refractive indexes, they are able to realize some novel ideas such as "perfect lens," which can image nanosized features surpassing the diffraction limit [31]. In addition, these submicrometer-structured metals have potential applications in ultratransmission imaging, light harvesting, optical waveguides, photodetectors, and nonlinear optics, etc. [6, 7, 35–37]. These discoveries have ignited great researching enthusiasms for metal array structures and the related properties.

In addition to the research areas described above, metallic structures can also be applied in some other areas including cancer treatment and optical storage [38–40]. In short, the studies on these state-of-the-art metals involve physics, materials, chemistry, biology, medicine, and many other disciplines, which provide great opportunities for scientific researches and practical applications.

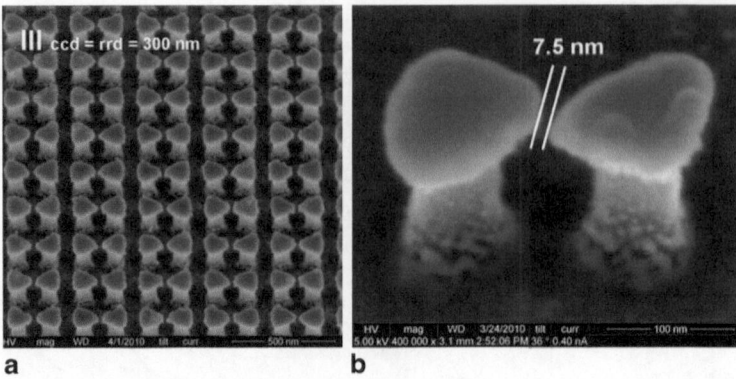

a   b

Fig. 1.2 Surface-enhanced Raman scattering (SERS) substrate fabricated via *top–down* approach. **a** Au bow-tie structure prepared via electron-beam lithography. **b** Close-up view of (**a**). (Reproduced with permission [41] Copyright 2010, American Chemical Society)

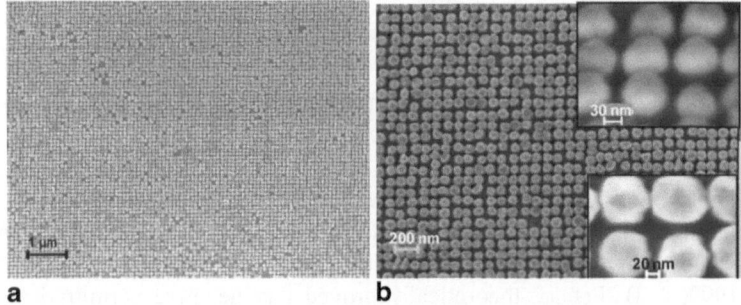

a   b

Fig. 1.3 Surface-enhanced Raman scattering (SERS) substrate fabricated via *bottom–up* approach. **a** Low-magnification scanning electron microscopy (SEM) image of 2D Ag nanoparticles. **b** Close-up view of (**a**). (Reproduced with permission [45] Copyright 2010, Wiley-VCH)

Driven by these prospects, scientists from a broad range of research fields have been seeking appropriate approaches to prepare these metal structures according to various uses. Nevertheless, none of the as-developed methods are fully satisfying at present. Since plasmonic devices usually have submicrometer textured surfaces, they are quite difficult to prepare. This situation will even get worse once hierarchical structures with complex morphologies are required. Present processing techniques within a "top–down" scheme (Fig. 1.2), e.g., conventional photolithography and focused ion beam (FIB) machining, suffer from a "shading effect." The applied light or mass flux can only machine surface in a line-of-sight way, yielding relatively simple textures [42]. Also, these methods are limited by complex processes, expensive equipments evolved, inefficiency, and product size. In comparison, "bottom-up" constructions have been developed to efficiently generate 3D structures (Fig. 1.3; [43–45]. Organic NPs can first self-assemble into specific 3D structures via some weak and noncovalent forces. These organic structures then serve as templates for subsequent metal deposition and will finally be removed to form their 3D metal replicas. The "bottom-up" routes, however, are limited by the achievable geometries of the 3D periodic structures and most of the as-prepared structures are close-packed. Therefore, lack in an efficient approach to 3D submicrometer metal structures is presently *the bottleneck* to be break through.

1.1.1 Functional Structures of Natural Species

It is quite interesting; however, that nature is good at generating 3D submicrometer structures. After millions of years of evolution, natural species have developed and optimized extremely elaborate hierarchical structures (biostructures) to survive the natural selection process. The features of these natural structures can range in size from nanometers to micrometers and their complexity is even beyond the finest human-designs. Such biostructures present living species a broad range of powerful functions in optics, electricity, acoustics, and heat, etc. [46–49].

Among these, optical structures might be the most important. Sunlight is not only the basis of vision [49, 50] but also the energy source for most of the species living on this planet. Many species have thus developed powerful and deliberate biostructures to utilize solar energy. Optical biostructures can be categorized into 1D optical gratings, 2D multilayer films, and 3D photonic structures [50–52].

Similar to human designs, 1D and 2D biostructures are usually optical gratings and multilayer films. Some flowers, like *Tulipa sp.* and *Hibiscus trionum*, exhibit shining colors owing to their photonic grating structures [53]. In addition, nacres have long been studied in biomimetics. These composites are composed of ceramic multilayer and can show iridescent colors via light interference on layers [54, 55]. Another example is insects' compound eyes. They are 2D periodic structures stacked by hexagonal units. Papilla array of compound eyes' corneal surface can eliminate light reflections and enhance the visual acuity of insects [56]. Such particular structure has been widely applied in artificial optics including huge inferred telescopes, micro-cameras, and fingerprint identification systems, etc. In comparison, some beetle exoskeletons and butterfly wing scales have brilliant colors generated by 3D photonic structures. These structures have periodically tuned dielectric coefficients in 3D space with spatial features comparable to visible light wavelengths [57, 58].

Gleaning these well-developed designs from nature, scientists can fabricate unique materials that are unachievable otherwise. Material scientists have replicated some key structures of natural species in desired components (i.e., target components). These target components include ceramics, metals, and polymers. The as-prepared materials can inherit original hierarchical biostructures with high fidelity and simultaneously have specific functions granted by the desired target components.

1.1.2   Replication of Biostructures into Functional Materials

Biostructures usually comprise organic and inorganic components with different proportions. They can accordingly be described as "inorganics" and "organics" depending on which components are dominant. There are two major differences between the replication of inorganic and organic structures. Apparently, the functional groups attached or exposed on the surfaces of these two types of templates are quite different. This may determine how the template surfaces should be pretreated. Also, the replication of these structures defer in a template-removal step. Acid treatments are usually applied to melt off inorganic templates, while calcinations under high temperatures are used to remove organic templates. Thus, various biostructures have to be converted to specific target components via different ways. More detailed fabrication strategies will be discussed in the next section. Herein, we just briefly introduce some as-prepared materials.

Inorganic biological skeleton mainly comprises $CaCO_3$, SiO_2, and some other ceramics. In 2000, F. C. Meldrum et al. coated Au on the skeletons of demosponge (*Ircinia oros*) and obtained their Au replicas via an acid treatment (see Fig. 1.4; [59]. In 2004, Y. H. Ha et al. prepared near-infrared 3D photonic structures with high dielectric coefficients using the skeletons of sea urchins as templates [60].

After that, many other inorganic biological structures were selected as biotemplates to generate state-of-the-art materials. For example, diatoms are a group of unicellular micro-organisms with over 200 genera and about 100,000 species all over the world. Their cell walls (frustules) are mainly made of SiO_2 and have hierarchical structures ranging in size from dozens to hundreds of nanometers. These

Fig. 1.4 Scanning electron microscopy (SEM) images of Au replicas of demosponge exoskeletons under **a** *low* and **b** *high* magnification. (Reproduced with permission [59] Copyright 2000, The Royal Society of Chemistry, UK)

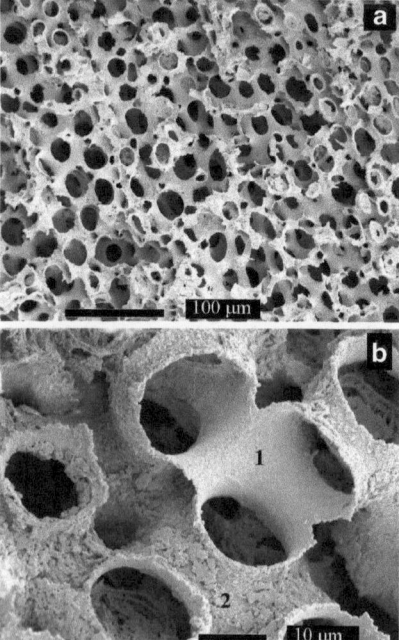

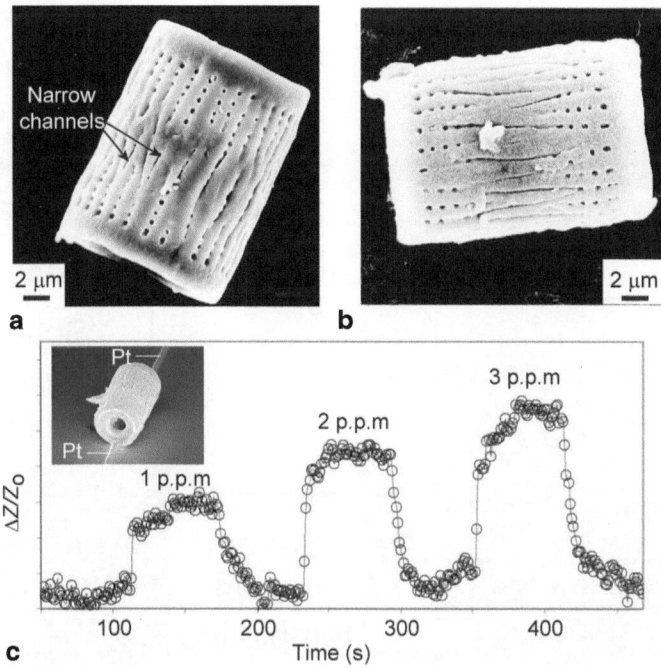

Fig. 1.5 **a** Natural diatom frustules. **b** Silicon replica. **c** Gas sensor based on (**b**). (Reproduced with permission [61] Copyright 2007, Nature Publishing Group)

structures are difficult to imitate either using "top–down" or "bottom-up" methods. However, one can get a variety of inorganic materials with such structures by replicating original frustules in target components. Z. H. Bao et al. chemically reduced the original silica frustules into Si in Mg vapors under 650 °C. These Si structures well-kept the frustules' original 3D structures down to nanoscale, and can be used as gas sensors, battery electrodes, and filters, etc. (Fig. 1.5; [61]).

Moreover, E. K. Payne et al. used a chemical vapor deposition (CVD) method to prepare Ag replicas of diatom frustules [63]. Therein, Ag was first deposited on frustule surface and the frustules were finally removed in a HF solution. These Ag replicas maintained the nanostructures of diatom frustules, and could enhance Raman signals of absorbed analytes [63]. Other nanostructured materials, such as Ag, C, SnO_2, TiO_2, ZrO_2, ZnS, Zn_2SiO_4, $BaTiO_3$, and $ZnFe_2O_4$, etc., have also successfully been prepared using diatom frustules as templates [64–72].

In addition to the inorganic biostructures discussed above, however, 80 % of biological structures are mainly composed of organic components. These structures can be found in species as small as bacteria, or as large as advanced plants and animals. Scientists have chosen different organic biostructures to get various functions. For example, as unicellular micro-organisms, bacteria have evolved a great number of submicrometer structures. Scientists have synthesized oxide and sulfide functional materials using bacteria as templates [73–76]. D. Zhang et al. converted *Streptococcus thermophilus* and *Lactobacillus bulgaricus* to oxide (ZnO and TiO_2) and sulfide (ZnS and PbS) hollow spheres and tubes. Excellent light-harvesting and

Fig. 1.6 Replication of photonic structures of *Lamprocyphus augustus* in SiO$_2$ and TiO$_2$ using a sol–gel approach. (Reproduced with permission [62] Copyright 2010, Wiley-VCH)

photocatalytic properties were achieved therein [73, 76, 77]. For plants, the hierarchical porous structures of their leaves and pollens were also gleaned to fabricate a series of functional oxides, which have promising applications in photocatalysis, light harvesting, and gas sensing, etc. [78–80]. Moreover, beetles are the largest group of animals with over 400,000 species. Most of their exoskeletons are of positive or inverse opal structures and exhibit typical structural colors. M. H. Bartl et al. used the green scales of *Lamprocyphus augustus* as templates to obtain 3D TiO$_2$ photonic structures with high dielectric coefficient, as shown in Fig. 1.6 [62].

1.2   Butterfly Scale Structures and Their Applications

Nevertheless, the photonic structures of beetles are buried under the exoskeleton surface. Such structures are regarded as "closed structures" as they may inhibit the infiltration processes of extrinsic solutions/vapors [62, 81]. Unfortunately, these processes are important to the replication since target components are usually transported by these solutions and vapors. Open biostructures are thus preferable to their closed counterparts. As typical open structures, wing scales of butterflies and moths have attracted considerable attention [46, 82–85]. These scales are mainly composed of chitin and have 3D photonic structures, which have been optimized by millions of years of natural selection and are far more intricate than artificial materials. Moreover, their scale microstructures are greatly diverse as butterflies and moths have more than 175,400 species (the second largest group in animals). It should also be noted that human beings have raised some butterflies and moths (e.g., silk moth) for thousands of years. The scales of butterflies and moths (referred hereafter as butterfly scales) can be collected after their natural deaths. Hence, the butterfly scales are appropriate natural biotemplates to replicate with regard to their abundance, open structures, and high variety in morphologies.

In addition to the advantages discussed above, butterfly scales are powerful functional units. Butterfly wings can present special colors under illumination. Some of these colors are structural and originate from scale textures. Typical scale structures responsible to colorization include so-called main ridges, ribs, and struts, etc. [86, 87], which will be introduced in Chap. 2. These structural units work together to effectively scatter, interfere, and diffract incident lights, leading to some significant optical phenomena that can even be observed by naked eyes.

In 1895, B. Waltar recorded his observations on *Morpho Menelaus*, a butterfly species with brilliant blue reflected along some special orientations. He immersed some wings of *M. Menelaus* into different liquids and found that the color changed [88]. After that, E. Merritt believed that structural colors are aroused by light interference between microlayers. This work established the interference theory on butterfly wings' structural colors [89]. With the development of modern technologies in optical measurement and structural characterization, much deeper understandings in the colorization mechanisms of butterfly scales have been achieved. For example, P. Vukusic et al. studied the colorization mechanisms of butterfly scales and indicated that they are natural 3D photonic structures [49]. They also presented theoretical explanations on the light scattering and absorption exerted by butterfly wings and compared the reflective properties between a single scale and a whole *Morpho* wing [46, 90]. S. Kinoshita and S. Yoshioka found that single scale of *Morpho didius* has anisotropy optical properties [91].

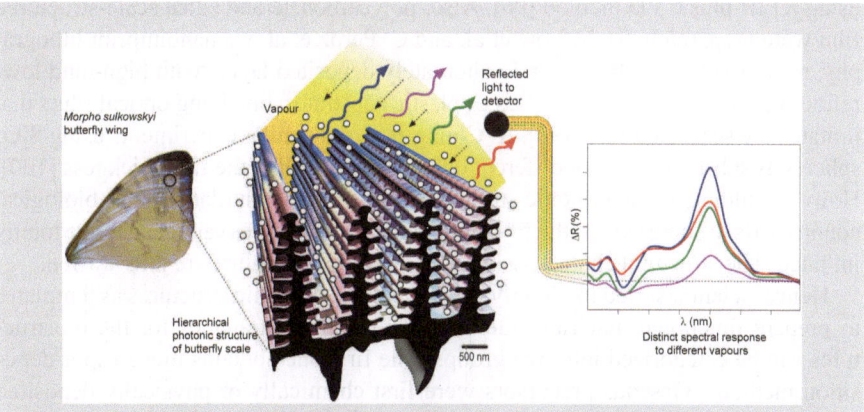

Box 1.2   Gas sensors based on Morpho butterfly wing. (Reproduced with permission [83] Copyright 2007, Nature Publishing Group)

Inspired by the photonic structures of butterflies, researchers explored the applications of original wings. R. A. Potyrailo et al. at GE Company found that the blue color of *Morpho* butterflies is closely related with surrounding atmospheres. They thus studied the gas sensing performance of wing scales [83]. Box 1.2 shows an experimental setup for gas detection. Once a wing was put in a certain atmosphere, corresponding gas molecules would quickly spread and fill into the gaps within the scale structure. Reflectance spectra could significantly change after a thin film of gas molecules fully covered the wing scale surfaces. This mechanism led to a new gas-sensing strategy with high sensitivity, high selectivity, and multichannels. Recently, they further designed a group of novel "butterfly infrared sensors" [92]. Therein, butterfly scales were combined with carbon nanotubes. When an infrared beam was cast onto the wing surface, the air filled in carbon nanotubes would be heated up. The air expansion and diffusion caused a deferring refractive index, which in turn changed the device color. Thus, the sensitivity of this sensor could increase with the amount of the loaded carbon nanotubes.

L. P. Biró et al. systematically studied the functions of butterfly scales as photonic crystals and gas sensors [93]. Recently, R. O. Prum et al. determined the 3D internal structures of butterfly wing scales using X-ray scattering techniques [94]. They even analyzed these results in views of biological evolution and cytology. In general, all these works have promoted the research on the functionalities of butterfly scales to a new level and provided a substantial basis for the generation of novel materials based on wing scales.

However, the main component of wing scales is chitin, which greatly limits the applications based on pure butterfly scales. For the past decade, great efforts have been spared to imitate the butterfly scale structures by artificial methods. In 2005, K. Watanabe et al. transformed diamond-like carbon into *Morph* butterfly structure using a FIB plus CVD method [96]. Also, polycarbonate and silica scale-structured film were prepared by H. Y. Low et al. and C. Peroz et al. via nanoimprint lithography, respectively [97, 98]. A. Satio alternately deposited layers with high- and low-refractive index onto polished glass or plastic substrates, obtaining optical films that imitated the scale structures [99]. K. Chung et al. did similar experiments using SiO_2 spheres as substrate, and got different colors by controlling the film thickness [100]. However, although some bionic structures topologically similar to their biological counterparts to some extent, the full hierarchical biostructures cannot so far be totally imitated. The structural characteristics well-designed by nature are thus wasted.

Hence, scientists tried to directly use butterflies' photonic structures as templates to prepare functional materials. Generally, replication strategies for the biostructures can be categorized into two groups. The first one contains those vapor deposition methods. Gaseous precursors were first chemically or physically deposited onto the template surface. Then, target components were either formed by calcinations at high temperatures or by chemical reactions conducted at mild temperatures.

These methods are highly controllable but suffered from so-called "shade effects." Since a mass deposition trace is determined by the line-of-sight nature, it might be difficult to achieve uniform coatings on surfaces with complex morphologies. Anyway, successful examples still existed. C. G. Spickermann et al. first deposited a layer of SiO_2 on the surface of butterfly wings using CVD, and then sintered the sample at 500 °C to remove the original organic templates [101]. Z. L. Wang et al. fabricated Al_2O_3 scale replicas with controlled layer thickness using atomic layer deposition (ALD). Experimental results suggested that these alumina structures can be used as light waveguides and splitters (Fig. 1.7; [95]). In 2009, A. Lakhtakia

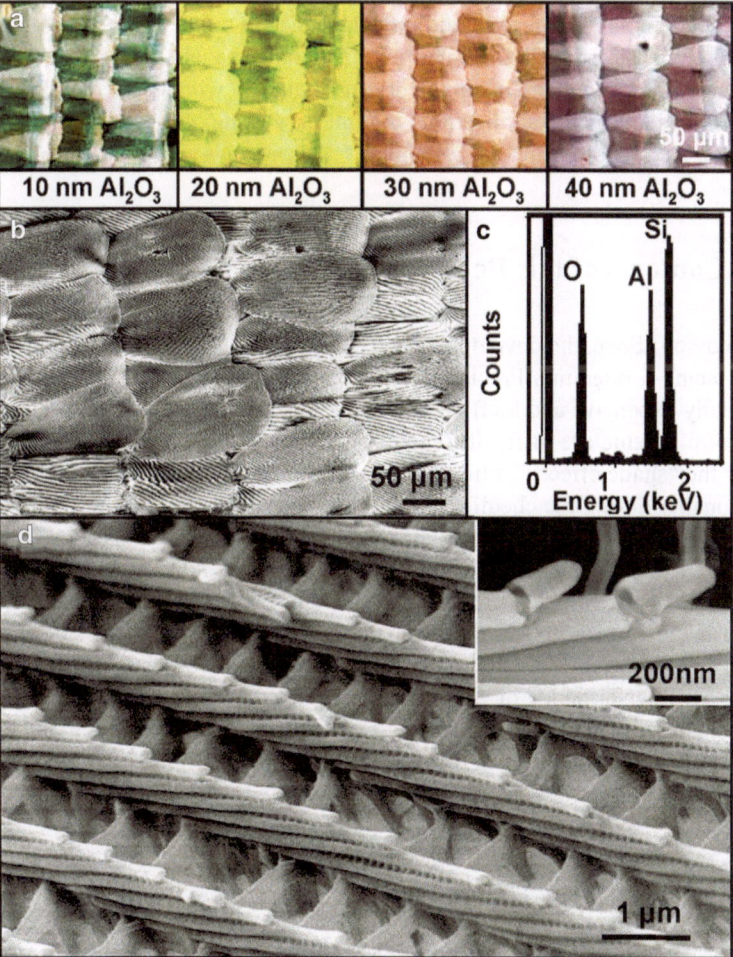

Fig. 1.7   As-synthesized Al_2O_3 replicas of *Morph* butterfly scales. **a** Optical microscopy image of scales with deferring Al_2O_3 coating thickness. **b** Low-magnification SEM image of Al_2O_3 replicas. **c** EDS results. **d** High-magnification SEM image of Al_2O_3 replicas. Inset shows the broken tips of rib structures, indicating that the original biostructures were removed by calcinations. (Reproduced with permission [95] Copyright 2006, American Chemical Society)

et al. used a conformal-evaporated-film-by-rotation (CEFR) technique to deposit GeSbSe on butterfly wings [102].

Instead, the other strategy, i.e., wet-chemical methods, might be more suitable to the conversion of biological templates to functional materials. In 2008, K. Sandhage et al. coated butterfly wings with precursors via a sol-gel method and subsequently removed organic templates at 450 °C to synthesize TiO_2 scale replicas [103]. They also got ternary compound ($BaTiO_3$) replicas using a similar method [104]. The samples basically maintained the structure original wing scales, while the nanosized rib structures were lost and the replicas shrunk after heat treatments. Other replicas in ZnO, ZrO_2, SnO_2, WO_3, and Fe_3O_4 were successfully prepared by D. Zhang et al. using various wet-chemical methods plus calcinations [105–111]. There were also many other studies that replicated butterfly scales in much more complex components like polydimethylsiloxane and lead zirconate titanate [112–116]. In general, these materials not only maintain the original butterfly wings' hierarchical structure but also have promising applications in light harvesting, gas sensing, and photo-catalyses, etc.

1.3   Contents of this Book

As discussed above, the development of submicrometer metallic structures are currently facing a dilemma. Precise fabrication methods within a top–down scheme are usually expensive and inefficient, while convenient synthesis methods within a bottom-up scheme are difficult to control. Moreover, top–down methods are limited by the "shade effects," while bottom-up approaches are affected by the loss in directionality during the chemical assembly process conducted at a submicrometer level. Hence, complicated 3D structures cannot be achieved using both schemes. To solve these problems, in this book we will introduce how natural butterfly wing scales with submicrometer structures can be chosen as templates and replicated in various metals. Three preparation strategies toward these butterfly-scale-structured metals will be introduced and compared in Chap. 2 and 3. Chapter 4 will demonstrate SERS performance of as-prepared metal replicas. In Chap. 5, the coupling between the scale structures and optical responses will be discussed. Finally, a brief perspective to this field will be provided at the end of this book.

References

1. Yablonovitch E (1987) Inhibited spontaneous emission in solid-state physics and electronics. Phys Rev Lett 58:2059–2062
2. Arpin KA, Mihi A, Johnson HT et al (2010) Multidimensional architectures for functional optical devices. Adv Mater 22:1084–1101
3. Barnes WL, Dereux A, Ebbesen TW (2003) Surface plasmon subwavelength optics. Nature 424:824–830

4. Pyayt AL, Wiley B, Xia Y et al (2008) Integration of photonic and silver nanowire plasmonic waveguides. Nat Nano 3:660–665
5. Anker JN, Hall WP, Lyandres O et al (2008) Biosensing with plasmonic nanosensors. Nat Mater 7:442–453
6. Atwater HA, Polman A (2010) Plasmonics for improved photovoltaic devices. Nat Mater 9:205–213
7. Wang W, Yang Q, Fan F et al (2011) Light propagation in curved silver nanowire plasmonic waveguides. Nano Lett 11:1603–1608
8. Sharma B, Frontiera RR, Henry A-I et al (2012) SERS: materials, applications, and the future. Mater Today 15:16–25
9. Henzie J, Lee J, Lee MH et al (2009) Nanofabrication of plasmonic structures. Annu Rev Phys Chem 60:147–165
10. Fang Y, Li Z, Huang Y et al (2010) Branched silver nanowires as controllable plasmon routers. Nano Lett 10:1950–1954
11. Noginov MA, Zhu G, Belgrave AM et al (2009) Demonstration of a spaser-based nanolaser. Nature 460:1110–1112
12. Park S, Won Hahn J (2009) Plasmonic data storage medium with metallic nano-aperture array embedded in dielectric material. Opt Express 17:20203–20210
13. Pala RA, White J, Barnard E et al (2009) Design of plasmonic thin-film solar cells with broadband absorption enhancements. Adv Mater 21:3504–3509
14. Falk AL, Koppens FHL, Yu CL et al (2009) Near-field electrical detection of optical plasmons and single-plasmon sources. Nature Phys 5:475–479
15. Kawata S, Ono A, Verma P (2008) Subwavelength colour imaging with a metallic nanolens. Nat Photon 2:438–442
16. Genet C, Ebbesen TW (2007) Light in tiny holes. Nature 445:39–46
17. Bozhevolnyi SI, Volkov VS, Devaux E et al (2006) Channel plasmon subwavelength waveguide components including interferometers and ring resonators. Nature 440:508–511
18. Benson O (2011) Assembly of hybrid photonic architectures from nanophotonic constituents. Nature 480:193–199
19. Chen QW, Li R, Zhang H et al (2011) Improved surface-enhanced Raman scattering on micro-scale Au hollow spheres: synthesis and application in detecting tetracycline. Analyst 136:2527–2532
20. Hu JA, Zhang CY (2010) Sensitive detection of nucleic acids with rolling circle amplification and surface-enhanced Raman scattering spectroscopy. Anal Chem 82:8991–8997
21. Spencer KM, Sylvia JM, Marren PJ et al (2004) Surface-enhanced Raman spectroscopy for homeland defense. P Soc Photo Opt Ins 5269:1–8
22. Zhou HB, Zhang ZP, Jiang CL et al (2011) Trinitrotoluene explosive lights up ultrahigh Raman scattering of nonresonant molecule on a top-closed silver nanotube array. Anal Chem 83:6913–6917
23. Schmuck C, Wich P, Kustner B et al (2007) Direct and label-free detection of solid-phase-bound compounds by using surface-enhanced Raman scattering microspectroscopy. Angew Chem Int Ed 46:4786–4789
24. Rana V, Canamares MV, Kubic T et al (2011) Surface-enhanced Raman spectroscopy for trace identification of controlled substances: morphine, codeine, and hydrocodone. J Forensic Sci 56:200–207
25. Fu Y, Lakowicz JR (2006) Enhanced fluorescence of Cy5-labeled DNA tethered to silver island films: fluorescence images and time-resolved studies using single-molecule spectroscopy. Anal Chem 78:6238–6245
26. Brouard D, Viger ML, Bracamonte AG et al (2011) Label-free biosensing based on multilayer fluorescent nanocomposites and a cationic polymeric transducer. ACS Nano 5:1888–1896
27. Peng H-I, Strohsahl CM, Leach KE et al (2009) Label-free DNA detection on nanostructured ag surfaces. ACS Nano 3:2265–2273
28. Staiano M, Matveeva EG, Rossi M et al (2009) Nanostructured silver-based surfaces: new emergent methodologies for an easy detection of analytes. ACS Appl Mater Interfaces 1:2909–2916

29. Hall WP, Anker JN, Lin Y et al (2008) A calcium-modulated plasmonic switch. J Am Chem Soc 130:5836–5837
30. Valentine J, Zhang S, Zentgraf T et al (2008) Three-dimensional optical metamaterial with a negative refractive index. Nature 455:376–379
31. Pendry JB (2000) Negative refraction makes a perfect lens. Phys Rev Lett 85:3966–3969
32. Shelby RA, Smith DR, Schultz S (2001) Experimental verification of a negative index of refraction. Science 292:77–79
33. Smith DR, Pendry JB, Wiltshire MC (2004) Metamaterials and negative refractive index. Science 305:788–792
34. Shalaev VM (2008) Transforming light. Science 322:384–386
35. Neutens P, Van Dorpe P, De Vlaminck I et al (2009) Electrical detection of confined gap plasmons in metal-insulator-metal waveguides. Nat Photon 3:283–286
36. Srituravanich W, Pan L, Wang Y et al (2008) Flying plasmonic lens in the near field for high-speed nanolithography. Nat Nanotech 3:733–737
37. Kim S, Jin J, Kim Y-J et al (2008) High-harmonic generation by resonant plasmon field enhancement. Nature 453:757–760
38. Challener WA, Peng C, Itagi AV et al (2009) Heat-assisted magnetic recording by a near-field transducer with efficient optical energy transfer. Nat Photon 3:220–224
39. Thomann I, Pinaud BA, Chen Z et al (2011) Plasmon enhanced solar-to-fuel energy conversion. Nano Lett 11:3440–3446
40. Au L, Zheng D, Zhou F et al (2008) A quantitative study on the photothermal effect of immuno gold nanocages targeted to breast cancer cells. ACS Nano 2:1645–1652
41. Hatab NA, Hsueh CH, Gaddis AL et al (2010) Free-standing optical gold bowtie nanoantenna with variable gap size for enhanced Raman spectroscopy. Nano Lett 10:4952–4955
42. Bao ZH, Ernst EM, Yoo S et al (2009) Syntheses of porous self-supporting metal-nanoparticle assemblies with 3D morphologies inherited from biosilica templates (diatom frustules). Adv Mater 21:474–478
43. Lim DK, Jeon KS, Hwang JH et al (2011) Highly uniform and reproducible surface-enhanced Raman scattering from DNA-tailorable nanoparticles with 1-nm interior gap. Nat Nanotechnol 6:452–460
44. Xia X, Zeng J, Mcdearmon B et al (2011) Silver nanocrystals with concave surfaces and their optical and surface-enhanced Raman scattering properties. Angew Chem Int Ed 50:12542–12546
45. Liberman V, Yilmaz C, Bloomstein TM et al (2010) A nanoparticle convective directed assembly process for the fabrication of periodic surface enhanced Raman spectroscopy substrates. Adv Mater 22:4298–4302
46. Vukusic P, Sambles JR, Lawrence CR (2000) Structural colour-colour mixing in wing scales of a butterfly. Nature 404:457–457
47. Vukusic P, Sambles JR (2003) Photonic structures in biology. Nature 424:852–855
48. Parker AR, Townley HE (2007) Biomimetics of photonic nanostructures. Nat Nanotechnol 2:347–353
49. Parker AR (2000) 515 million years of structural colour. J Opt A Pure Appl Op 2:R15–R28
50. Parker AR (2004) A vision for natural photonics. Philos T Roy Soc A 362:2709–2720
51. Biro LP, Vigneron JP (2011) Photonic nanoarchitectures in butterflies and beetles: valuable sources for bioinspiration. Laser Photonics Rev 5:27–51
52. Fan TX, Chow SK, Di Z (2009) Biomorphic mineralization: from biology to materials. Prog Mater Sci 54:542–659
53. Whitney HM, Kolle M, Andrew P et al (2009) Floral iridescence, produced by diffractive optics, acts as a cue for animal pollinators. Science 323:130–133
54. Zhang GS, Huang ZQ (2010) Two-dimensional amorphous photonic structure in the ligament of bivalve *Lutraria maximum*. Opt Express 18:13361–13367
55. Tan T, Wong D, Lee P (2004) Iridescence of a shell of mollusk *Haliotis glabra*. Opt Express 12:4847–4854

56. Huang J, Wang X, Wang ZL (2008) Bio-inspired fabrication of antireflection nanostructures by replicating fly eyes. Nanotechnology 19:025602
57. Parker AR, Welch VL, Driver D et al (2003) Structural colour-opal analogue discovered in a weevil. Nature 426:786–787
58. Smith GS (2009) Structural color of *Morpho* butterflies. Am J Phys 77:1010–1019
59. Meldrum FC, Seshadri R (2000) Porous gold structures through templating by echinoid skeletal plates. Chem Commun 29–30
60. Ha YH, Vaia RA, Lynn WF et al (2004) Three-dimensional network photonic crystals via cyclic size reduction/infiltration of sea urchin exoskeleton. Adv Mater 16:1091–1094
61. Bao ZH, Weatherspoon MR, Shian S et al (2007) Chemical reduction of three-dimensional silica micro-assemblies into microporous silicon replicas. Nature 446:172–175
62. Galusha JW, Jorgensen MR, Bartl MH (2010) Diamond-structured titania photonic-bandgap crystals from biological templates. Adv Mater 22:107–110
63. Payne EK, Rosi NL, Xue C et al (2005) Sacrificial biological templates for the formation of nanostructured metallic microshells. Angew Chem Int Ed 44:5064–5067
64. Zhou H, Fan TX, Li XF et al (2009) Bio-inspired bottom-up assembly of diatom-templated ordered porous metal chalcogenide meso/nanostructures. Eur J Inorg Chem 2009:211–215
65. Losic D, Mitchell JG, Voelcker NH (2005) Complex gold nanostructures derived by templating from diatom frustules. Chem Commun 4905–4907
66. Losic D, Triani G, Evans PJ et al (2006) Controlled pore structure modification of diatoms by atomic layer deposition of TiO_2. J Mater Chem 16:4029–4034
67. Holmes SM, Graniel-Garcia BE, Foran P et al (2006) A novel porous carbon based on diatomaceous earth. Chem Commun 2662–2663
68. Weatherspoon MR, Dickerson MB, Wang G et al (2007) Thin, conformal, and continuous SnO_2 coatings on three-dimensional biosilica templates through hydroxy-group amplification and layer-by-layer alkoxide deposition. Angew Chem Int Ed 46:5724–5727
69. Cai Y, Dickerson MB, Haluska MS et al (2007) Manganese-doped zinc orthosilicate-bearing phosphor microparticles with controlled three-dimensional shapes derived from diatom frustules. J Am Ceram Soc 90:1304–1308
70. Liu Z, Fan T, Zhou H et al (2007) Synthesis of $ZnFe_2O_4/SiO_2$ composites derived from a diatomite template. Bioinspir Biomim 2:30–35
71. Shian S, Cai Y, Weatherspoon MR et al (2006) Three-dimensional assemblies of zirconia nanocrystals via shape-preserving reactive conversion of diatom microshells. J Am Ceram Soc 89:694–698
72. Cai Y, Sandhage KH (2005) Zn_2SiO_4-coated microparticles with biologically-controlled 3D shapes. Phys Status Solidi A 202:R105–R107
73. Zhou H, Fan T, Ding J et al (2012) Bacteria-directed construction of hollow TiO_2 micro/nanostructures with enhanced photocatalytic hydrogen evolution activity. Opt Express 20(Suppl 2):A340–A350
74. Zhang TJ, Wang W, Zhang DY et al (2010) Biotemplated synthesis of gold nanoparticle-bacteria cellulose nanofiber nanocomposites and their application in biosensing. Adv Funct Mater 20:1152–1160
75. Alloyeau D, Stephanidis B, Zhao X et al (2011) Biotemplated synthesis of metallic nanoclusters organized in tunable two-dimensional superlattices. J Phys Chem C 115:20926–20930
76. Zhou H, Fan T, Zhang D et al (2007) Novel bacteria-templated sonochemical route for the in situ one-step synthesis of ZnS hollow nanostructures. Chem Mater 19:2144–2146
77. Zhou H, Fan T, Han T et al (2009) Bacteria-based controlled assembly of metal chalcogenide hollow nanostructures with enhanced light-harvesting and photocatalytic properties. Nanotechnology 20:085603
78. Li X, Fan T, Zhou H et al (2009) Enhanced light-harvesting and photocatalytic properties in morph-TiO_2 from green-leaf biotemplates. Adv Funct Mater 19:45–56
79. Zhou H, Li X, Fan T et al (2010) Artificial inorganic leafs for efficient photochemical hydrogen production inspired by natural photosynthesis. Adv Mater 22:951–956

80.   Song F, Su H, Han J et al (2012) Bioinspired hierarchical tin oxide scaffolds for enhanced gas sensing properties. J Phys Chem C 116:10274–10281

81.   Jorgensen MR, Bartl MH (2011) Biotemplating routes to three-dimensional photonic crystals. J Mater Chem 21:10583–10591

82.   Sweeney A, Jiggins C, Johnsen S (2003) Insect communication: Polarized light as a butterfly mating signal. Nature 423:31–32

83.   Potyrailo RA, Ghiradella H, Vertiatchikh A et al (2007) Morpho butterfly wing scales demonstrate highly selective vapour response. Nat Photon 1:123–128

84.   Vukusic P, Sambles JR, Lawrence CR (2004) Structurally assisted blackness in butterfly scales. Proc Biol Sci Roy Soc 271 (Suppl 4):4S237–S239

85.   Vukusic P, Hooper I (2005) Directionally controlled fluorescence emission in butterflies. Science 310:1151–1151

86.   Tan Y, Gu J, Zang X et al (2011) Versatile fabrication of intact three-dimensional metallic butterfly wing scales with hierarchical sub-micrometer structures. Angew Chem Int Ed 50:8307–8311

87.   Tan YW, Gu JJ, Xu LH et al (2012) High-density hotspots engineered by naturally piled-up subwavelength structures in three-dimensional copper butterfly wing scales for surface-enhanced Raman scattering detection. Adv Funct Mater 22:1578–1585

88.   Walter B (1895) Die oberflaehen-oder sehillerfarben. F. Vieweg und Sohn, Braunsehweig

89.   Merritt E (1925) A spectrophotometric study of certain cases of structural color. J Opt Soc Am 11:93–97

90.   Vukusic P, Sambles JR, Lawrence CR et al (1999) Quantified interference and diffraction in single Morpho butterfly scales. Proc Roy Soc Lond B 266:1403–1411

91.   Yoshioka S, Kinoshita S (2006) Single-scale spectroscopy of structurally colored butterflies: measurements of quantified reflectance and transmittance. J Opt Soc Am A 23:134–141

92.   Pris AD, Utturkar Y, Surman C et al (2012) Towards high-speed imaging of infrared photons with bio-inspired nanoarchitectures. Nat Photon 6:195–200

93.   Biro LP, Balint Z, Kertesz K et al (2003) Role of photonic-crystal-type structures in the thermal regulation of a lycaenid butterfly sister species pair. Phys Rev E 67:021907

94.   Saranathan V, Osuji CO, Mochrie SGJ et al (2010) Structure, function, and self-assembly of single network gyroid (I4132) photonic crystals in butterfly wing scales. Proc Natl Acad Sci USA 107:11676–11681

95.   Huang J, Wang X, Wang ZL (2006) Controlled replication of butterfly wings for achieving tunable photonic properties. Nano Lett 6:2325–2331

96.   Watanabe K, Hoshino T, Kanda K et al (2005) Brilliant blue observation from a Morpho-butterfly-scale quasi-structure. Jpn J Appl Phys 2(44):L48–L50

97.   Kustandi TS, Low HY, Teng JH et al (2009) Mimicking domino-like photonic nanostructures on butterfly wings. Small 5:574–578

98.   Saison T, Peroz C, Chauveau V et al (2008) Replication of butterfly wing and natural lotus leaf structures by nanoimprint on silica sol-gel films. Bioinspir Biomim 3:046004

99.   Saito A, Miyamura Y, Ishikawa Y et al (2009) Reproduction, mass-production, and control of the Morpho-butterfly's blue. In: Suleski TJ, Schoenfeld WV, (eds) Advanced fabrication technologies for micro/nanooptics and photonics II. SPIE, p 720506. Bellingham, USA

100. Chung K, Yu S, Heo C-J et al (2012) Angle-independent reflectors: flexible, angle-independent, structural color reflectors inspired by Morpho butterfly wings. Adv Mater 24:2366–2366

101. Cook G, Timms PL, Goltner-Spickermann C (2003) Exact replication of biological structures by chemical vapor deposition of silica. Angew Chem Int Ed 42:557–559

102. Lakhtakia A, Martin-Palma RJ, Motyka MA et al (2009) Fabrication of free-standing replicas of fragile, laminar, chitinous biotemplates. Bioinspir Biomim 4:034001

103. Weatherspoon MR, Cai Y, Crne M et al (2008) 3D rutile titania-based structures with Morpho butterfly wing scale morphologies. Angew Chem Int Ed 47:7921–7923

104. Vernon JP, Fang YN, Cai Y et al (2010) Morphology-preserving conversion of a 3D bioorganic template into a nanocrystalline multicomponent oxide compound. Angew Chem Int Ed 49:7765–7768
105. Zhang W, Zhang D, Fan TX et al (2006) Biomimetic zinc oxide replica with structural color using butterfly (*Ideopsis similis*) wings as templates. Bioinspir Biomim 1:89–95
106. Zhang W, Zhang D, Fan TX et al (2009) Novel photoanode structure templated from butterfly wing scales. Chem Mater 21:33–40
107. Chen Y, Gu JJ, Zhang D et al (2011) Tunable three-dimensional ZrO_2 photonic crystals replicated from single butterfly wing scales. J Mater Chem 21:15237–15243
108. Peng WH, Zhu SM, Wang WL et al (2012) 3D network magnetophotonic crystals fabricated on *Morpho* butterfly wing templates. Adv Funct Mater 22:2072–2080
109. Zhu S, Liu X, Chen Z et al (2010) Synthesis of Cu-doped WO_3 materials with photonic structures for high performance sensors. J Mater Chem 20:9126–9132
110. Song F, Su HL, Chen JJ et al (2011) Bioinspired ultraviolet reflective photonic structures derived from butterfly wings (*Euploea*). Appl Phy Lett 99:163705
111. Song F, Su H, Han J et al (2009) Fabrication and good ethanol sensing of biomorphic SnO_2 with architecture hierarchy of butterfly wings. Nanotechnology 20:495502
112. Silver J, Withnall R, Ireland TG et al (2005) Novel nano-structured phosphor materials cast from natural *Morpho* butterfly scales. J Mod Optic 52:999–1007
113. Ji-Zhong Z, Zhong-Ze G, Hai-Hua C et al (2006) Inverse *Mopho* butterfly: a new approach to photonic crystal. J Nanosci Nanotech 6:1173–1176
114. Xu Z, Yu K, Li B et al (2011) Optical properties of SiO_2 and ZnO nanostructured replicas of butterfly wing scales. Nano Res 4:737–745
115. Kang SH, Tai TY, Fang TH (2010) Replication of butterfly wing microstructures using molding lithography. Curr Appl Phys 10:625–630
116. Li B, Zhou J, Zong R et al (2006) Ordered ceramic microstructures from butterfly biotemplate. J Am Ceram Soc 89:2298–2300

Chapter 2
Toward Metallic Butterfly Wing Scales

2.1 Introduction

Till now, natural biostructures have already been converted to a broad range of oxides. To prepare these replicas, metal ions were first coordinated on the surface of biological templates via an impregnation process. The hybrids were subsequently sintered in *air* under high temperatures to form desired oxides, with the original biological skeletons simultaneously removed [1–5].

However, preparing metallic structures based on biotemplates is a rising field. Metal nanoparticles (NPs) have comparatively high-surface energies and low-melting points [6–10]. These natures can cause severe particle aggregation and growth during the sintering process. Additionally, since some metals are prone to oxidation, such an approach of impregnation plus calcinations in air is unsuitable to preparing hierarchically structured metals. Researchers thus coated metal particles onto biostructures using physical vapor deposition (PVD) and chemical vapor deposition (CVD) [11, 12]. Nevertheless, the line-of-sight nature of these vapor deposition methods prevented a complete inheritance of the original three-dimensional (3D) biomorphologies [13].

Hence, all the approaches introduced in this book are based on wet-chemical methods. Some routes were conducted under high temperatures (e.g., H_2 reduction) while others were at mild temperatures (e.g., photoreduction). In this chapter, two strategies including H_2 reduction and photoreduction will be introduced. Experimental parameters, degrees of morphology preserving, and related mechanisms will be discussed in details to evaluate these two methods.

We select copper (Cu) and silver (Ag) as two target components. These metals have been widely applied due to their good electrical and thermal properties. Presently, different preparation routes for Ag and Cu materials with subwavelength features have been widely reported [13–19]. These methods include calcinations in reducing atmospheres and chemical syntheses conducted under mild temperatures. For example, Z. H Bao et al. first converted the bio-SiO_2 of diatom cell walls to Si by application of Mg steam under 650 °C [13, 19]. Then, Ag, Au, and Pd were deposited onto the Si diatoms. The as-prepared metals well-kept the 3D characteristics

J. Gu et al., *Metallic Butterfly Wing Scales*, SpringerBriefs in Materials,
DOI 10.1007/978-3-319-12535-0_2

of original biological structures down to a nanometer scale [13]. Therefore, this work has a certain reference for preparing metals with submicrometer superstructures gleaned from butterfly wing scales.

2.2 Biotemplate Selection

Lepidopterans (butterflies and moths) have ~174,500 identified species, and their colorizations may be achieved by two ways [20]. One results from wing pigments. Colors generated in this way are also known as pigment colors or chemical colors. The other is aroused by specific optical reflections, which is mainly determined by the interactions between submicrometer-textured wing scales and incident lights. Colors induced by this process are also known as structural colors [21, 22]. For butterfly wings, typical structural colors are usually shining blue and green, while pigment colors are mainly black, white, red, and yellow, etc. In many cases, these two colorization mechanisms can coexist on butterfly wings. Figure 2.1 shows some species of butterflies and moths. Wing colors in Fig. 2.1a–e are structural, while those in f–l are caused by pigments.

Fig. 2.1 Typical species of butterfly and moth

Fig. 2.2 Scanning electron microscopy (SEM) images of different kinds of butterfly and moth scales are shown in Fig. 2.1. Insets are the corresponding optical microscopy images. Scale bars: 2 μm. (50 μm for inset)

Surface textures of wings exhibited in Fig. 2.1a–e are shown in Fig. 2.2. These scanning electron microscopy (SEM) images were taken from the rectangular areas marked in Fig. 2.1. As shown in Fig. 2.2, the surfaces of butterfly wings are covered with numerous small and flat scales, with a typical size of 100–200 μm in length and 50–100 μm in width. Different scale structures result in deferring structural colors shown in the insets of Fig. 2.2a–e, which are in association with different biological functions. The morphology diversity in wing scales presents researchers enough freedom in selecting appropriate butterfly wing structures as biotemplates to fulfill specific functions.

Herein, two species of butterflies (*Euploea mulciber* and *Papilio paris*) with different microstructures are selected as biotemplates to replicate. These butterflies are widely distributed and easy to collect. The forewing scales of *E. mulciber* have a 3D photonic structure, while those of *P. paris* have a honeycomb structure.

Figure 2.3 shows the digital photo and optical microscopy image of an *E. mulciber* (*Cramer*) butterfly. The dorsal side of its forewing exhibits blue-violet with a metallic luster. When the butterfly dances, its wings may reflect dull purple, bright purple, and bright blue of sunlight in different directions. Previous studies indicated

Fig. 2.3 *E. mulciber.*
a Whole wing. **b** Forewing
of **a**. **c** Optical microscopy
image of wing scales on the
dorsal side of **b**

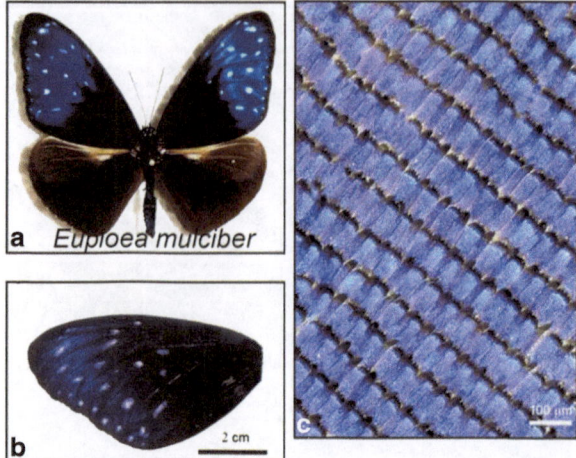

Fig. 2.4 Scanning electron
microscopy (SEM) images
of blue scales on the dorsal
side of *E. mulciber* forewing. Structural features are
denoted in **c** and **d**. (Reproduced with permission [24].
Copyright 2013, American
Chemical Society)

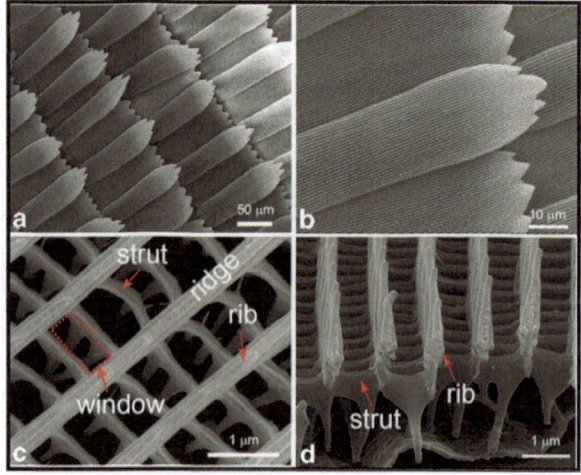

that such blue-violet colors come from the photonic structures of scales rather than pigments [23]. These tiling scales are shown in Fig. 2.3c. In comparison, the colors of *E. mulciber* hindwings do not change with the observation directions, suggesting that the brown colors come from pigments.

The surface morphology of individual single scale was further observed using SEM. Since original wings are not conductive, Au sputtering is needed to treat the wing surface before SEM observations. As shown in Fig. 2.4a, the wing surface is uniformly covered by many scales that are orderly arranged in a certain direction, just like tiles on a roof. The size of an individual scale is about 50×100 μm.

Under higher magnifications, it can be observed that these scales have complex microstructures. There are parallel main ridges with a period of ~ 1 μm and short

Fig. 2.5 *P. paris*. **a** Whole butterfly. **b** Forewing of (**a**). **c** Optical microscopy image of wing scales on the dorsal side of (**b**)

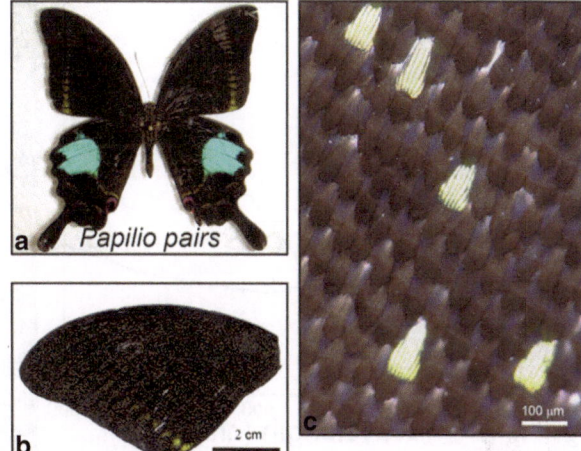

Fig. 2.6 Scanning electron microscopy (SEM) images of the black scales on the dorsal side of *P. paris* forewing. **c** and **d** are the top-view and side-view SEM images of (**b**), respectively

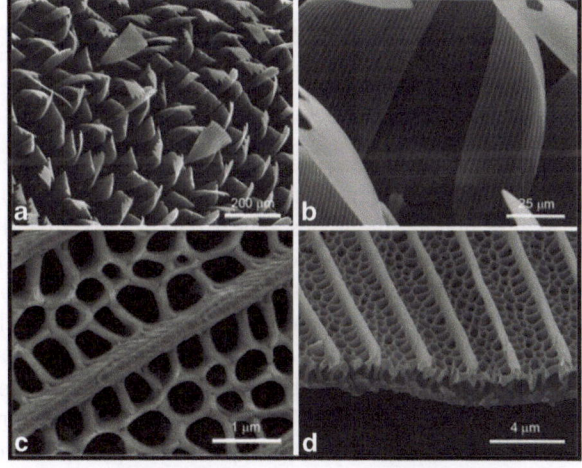

struts vertical to main ridges, forming the "window" structures marked in Fig. 2.4c. Figure 2.4d shows a cross-sectional view of the scale. Overlapped thin rib structures locate on each individual ridge, with a certain angle against the scale surface. The spacing between adjacent ribs is about 100–200 nm.

The digital photo and optical image of a *P. paris* butterfly are shown in Fig. 2.5. Since its hindwing contains various types of scales, we mainly deal with its forewing (Fig. 2.5b). The dorsal side of the forewing is mainly covered by black scales, with some bright yellow/green scales sparsely distributed within. These yellow/green scales have different contours and shapes to their black counterparts (Fig. 2.5c).

The structure of black scales of *P. paris* were studied with SEM. As shown in Fig. 2.6, fine microstructures of the black scale can be observed. The spacing between two adjacent ridges is about 2.5 μm. Disorderly arranged holes with several

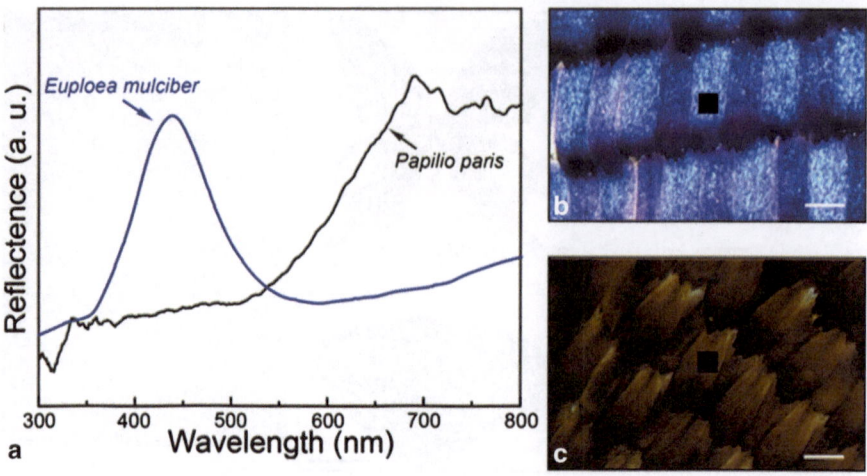

Fig. 2.7 a Reflectance spectra of two natural butterfly wing scales with different microstructures.
(**b**) and (**c**) are optical microscopy images of **b** the blue scales of *E. mulciber* and **c** the black scales
of *P. paris*. The *black rectangles* denote the selected areas for spectra collection. Scale bar: 50 μm.
To prepare these samples, the original butterfly wings were first washed in alcohol for 20 min and
dried in air. The treated wings were then placed on a black substrate for measurement. The reflectance spectra was recorded with vertical incidence and reflection

hundred nanometers in diameter are distributed between main ridges (Fig. 2.6c).
This "quasi-honeycomb" structure is beneficial to the multiple scatterings of light
in scales, making the scales appear darker [25]. In addition, there are no other ordered periodic structures except main ridges (Fig. 2.6c), which might explain why
structural colors are not observed on this type of scales.

Both scale structures are important for the survival of the corresponding butterflies and are hard to be manually reproduced. The scales of *E. mulciber* make its
forewing surface shine with a blue luster, which is beneficial to courtship and in-
traspecies communication. In comparison, the quasi-honeycomb structures of black
scales can help *P. paris* to absorb solar energy much more effectively, which helps
to balance the heat of butterflies [26, 27]. Figure 2.7 shows visible reflectance spectra of the forewings of *E. mulciber* and *P. paris*, respectively. For *E. mulciber* scales,
there is a strong reflection peak at ~450 nm, while no reflection peak can be found
in reflectance spectra of *P. paris*.

2.3   Cu Scale Replicas Prepared via H_2 Reduction Under High Temperatures

The first route toward metal scale replicas is via H_2 reduction. This strategy can
be divided into two steps. First, specific oxide scales as intermediates are prepared
via calcinations. Then, these oxide scales are reduced in H_2 atmosphere to form

metal replicas. Cu is selected herein as a target component for demonstration. This method can be extended to prepare other transition metals that can be reduced by H$_2$.

2.3.1   Sample Preparation

The CuO butterfly wings can be fabricated using an impregnation method described in Ref. [3]. A precursor with Cu^{2+} ions was first prepared. Twenty gram Cu(NO$_3$)$_2$ 5H$_2$O was dissolved in 50 mL ethanol and stirred for an hour. Ethylene glycol was then added into the Cu(NO$_3$)$_2$ ethanol solution to form a precursor. Butterfly wings pretreated using the method reported in Ref. [3] were then immersed in this precursor for 24 h. After that, the wings were taken out, rinsed in deionized water, placed on a silicon wafer, and dried in air for 12 h. The wings were then moved into a muffle furnace and heated up to 500 °C with a heating rate of 1 °C/min to remove the biological wing skeletons. CuO butterfly scales were thus obtained.

These CuO scales were then reduced in a H$_2$ atmosphere to obtain Cu wing replicas. To achieve this, CuO wings were put in a furnace with H$_2$ atmosphere. Mixed flowing gas of hydrogen/argon (volume ratio of 4:1) was adopted therein, with a gas flow rate of 200 mL/min and a sintering temperature of 250 °C. After 1 h, the Cu butterfly wings were fabricated.

2.3.2   Characterizations

Figure 2.8a and b show SEM images of an as-prepared CuO scale. This CuO scale has a spatial periodic structure that retains the original ridges and the periodic window structures (see Fig. 2.4c). Figure 2.8c shows an energy-dispersive X-ray spectroscopy (EDS) result on the CuO scale. This spectrum suggests that the sample mainly contained Cu, O, and Si, while the detected Si signal originated from the Si wafer. X-ray diffraction (XRD) analysis (Fig. 2.8d) indicates that the synthesized film was mainly composed of crystallized CuO. Based on the Scherrer equation, the CuO grain size was estimated to be ∼ 12 nm by measuring the half-width of the (002) diffraction peak.

Transmission electron microscopy (TEM) analyses on the obtained CuO butterfly wings are provided in Fig. 2.9. Figure 2.9c reveals the lattice fringes of a single NP with a lattice spacing of ∼0.252 nm, which corresponds to the (002) plane of CuO. Corresponding selected area electron diffraction (SAED) pattern is shown in Fig. 2.9d. The diffraction rings are well consistent with those of CuO and can be indexed as (110), (002), (200), ($\bar{2}$02), and (020), respectively.

Figure 2.10 compares a CuO wing with its Cu replica obtained via H$_2$ reduction. The CuO wing exhibits an iridescent characteristic, suggesting that the original photonic biostructures were well-preserved. In comparison, the sample color became

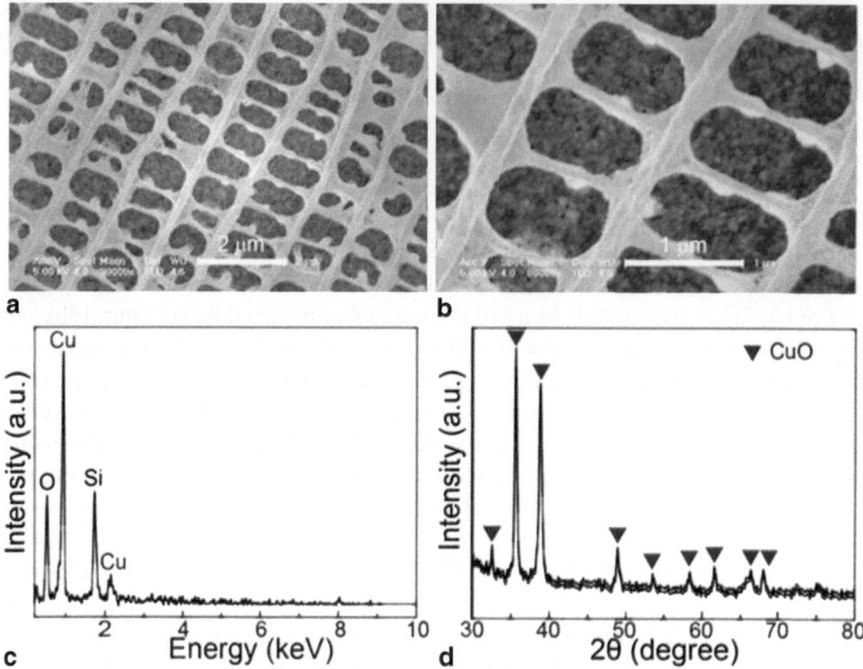

Fig. 2.8 a and **b** scanning electron microscopy (SEM) images of CuO replicas. **c** Energy-dispersive X-ray (EDS) result of CuO replicas. **d** X-ray diffraction (XRD) result of CuO replicas

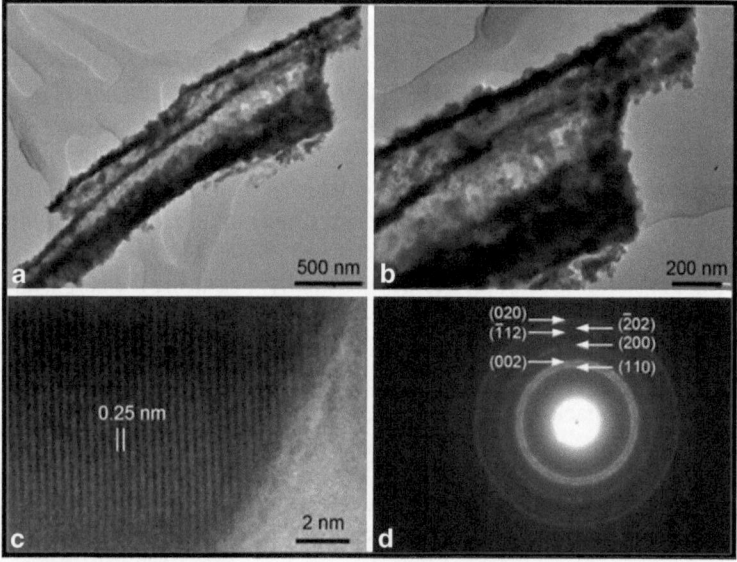

Fig. 2.9 a and **b** transmission electron microscopy (TEM) images of CuO replicas. **c** Lattice fringes of a CuO NP. **d** Selected area electron diffraction (SAED) pattern of the corresponding areas

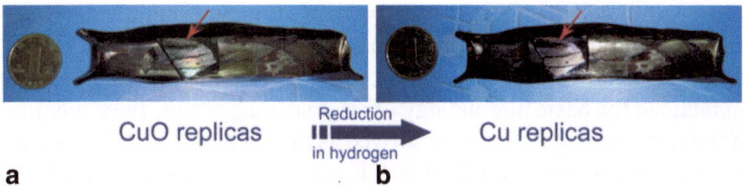

Fig. 2.10 Optical images of **a** CuO and **b** Cu replicas

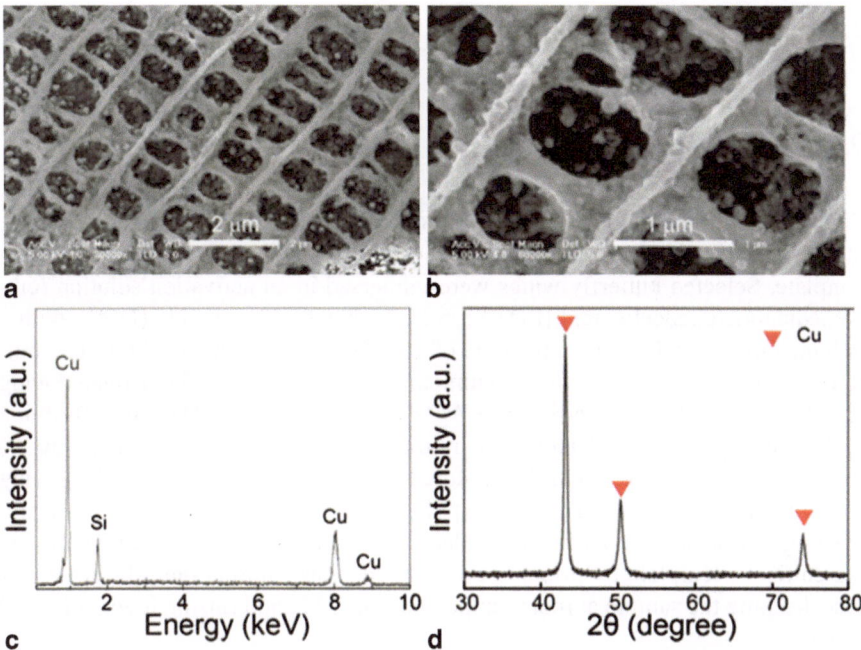

Fig. 2.11 **a** and **b** scanning electron microscopy (SEM) images of Cu replicas prepared by the reduction of CuO intermediates in H$_2$ at 250 °C. **c** Energy-dispersive X-ray spectroscopy (EDS) result of Cu replicas. **d** X-ray diffraction (XRD) result of Cu replicas

dark red after the calcination in H$_2$, indicating that the CuO has been reduced to metal Cu.

Figure 2.11a and b show the SEM images of an as-reduced Cu scale. As shown in this figure, the Cu scale basically retains the original morphology of wing scales (e.g., main ridges and window structures). However, finer scale structures are not well preserved. Compared with the CuO scale, the surface of the Cu replica becomes rough. Cu NPs grow larger and some parts of the ridges even collapse. Figure 2.11c gives the EDS results of the Cu replicas, which indicates that the sample only contains Cu and Si that originates from the Si substrate. XRD results (Fig. 2.11d) indicate that the synthesized film is crystallized Cu. Using the Scherrer equation, the average grain size of Cu is calculated to be ~30 nm.

These results confirm that CuO scales with the structures inherited from butterfly scales may be obtained with the impregnation plus calcination route. These scales as intermediates can be reduced into Cu in H_2 atmosphere. Although the reduced Cu film retained the basic morphology of original wing scales, finer submicrometer superstructures of the scales were destroyed as the Cu grains rapidly grew up under the high-reduction temperature. Hence, it is necessary to find an alternative method to prepare metal butterfly wings under mild temperatures.

2.4   Ag Scale Replicas Prepared via Photoreduction Under Mild Temperatures

2.4.1   Sample Preparation

Figure 2.12 illustrates the route towards Ag scale replicas via photoreduction. The method described in Ref. [28] was first used to activate the original butterfly wing template. Selected butterfly wings were immersed in an activation solution (ethylenediaminetetraacetic acid (EDTA) in N, N-dimethylformamide (DMF) with a volume ratio of 1:10) at a temperature of 110 °C for several hours. After that, butterfly wings were washed with an ammonia/ethanol solution. The activated wing templates were then immersed in a $AgNO_3$ water solution. The butterfly wing scales with $AgNO_3$ loaded upon were irradiated under an ultraviolet source (30 W) for 12 h in a closed box. After the photoreduction process, composites of Ag/wing were obtained. A fine needle with a tip diameter of ~ 30 µm was used to transfer the single wing scale to a Si wafer under a stereomicroscope. We finally removed the chitinous template by dripping a H_3PO_4 aqueous solution onto the scale surface, keeping the sample at room temperature for 72 h, and rinsing it with distilled water.

2.4.2   Formation Mechanisms

The original butterfly wing is mainly composed of chitin and a small amount of proteins [29]. Chitin is a polymer of 2-(acetylamino)-2-deoxy-D-glucose ($C_8H_{13}NO_5$).

Fig. 2.12 Fabrication processes for Ag butterfly wing scales

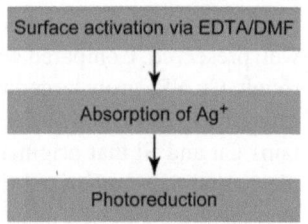

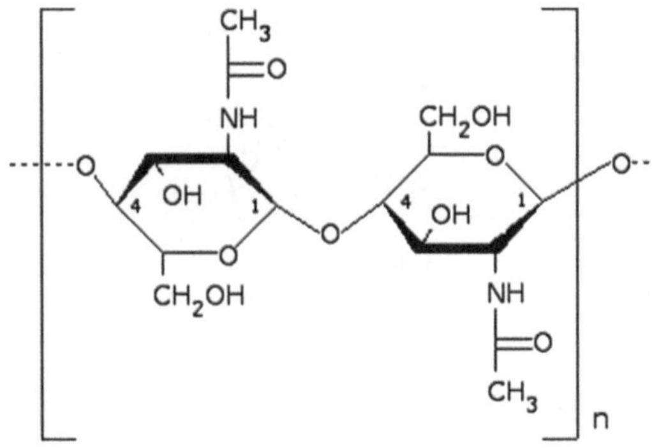

Fig. 2.13 Structural formula
of chitin

Its molecular structure is shown in Fig. 2.13. As a linear polysaccharide polymer, chitin has a portion of naturally exposed amino, hydroxyl, and carboxyl groups [29]. But these intrinsic active cites are not enough to coordinate metal ions. To get a continuous metal coating, extrinsic amino and carboxyl groups should be first introduced onto the wing surface via proper pretreatments. This process may help to improve the tendency of the wing surface to absorb and coordinate metal ions [30].

Figure 2.14 shows the Fourier transform infrared spectroscopy (FTIR) spectra of the wing templates before and after pretreatments. The peaks located at 1655 and 1543 cm^{-1} correspond to amide I and amide II absorption band from proteins. 1157, 1115, 1074, and 1543 cm^{-1} peaks result from characteristic vibration modes of C–O in chitin [31]. The 1716 cm^{-1} absorption band corresponds to the stretching vibration modes of –COOH from the residues of aspartic and glutamic acids. Bands at 3250 and 3450 cm^{-1} correspond to the characteristic peaks of the N–H (amide II and primary amino) and the stretching vibration of –OH in chitin. Thus, there be some active sites (e.g., –COOH, –NH$_2$, and –OH) beneficial for metal coordination on the scale surfaces. However, additional activation treatments are still needed to further increase the number and obtain a better distribution of the active sites. After the activation treatment in EDTA/DMF solution (Fig. 2.14), the absorption intensity at 3444 cm^{-1} slightly increased and a noticeable double-peak occurred in the FTIR spectrum. This is due to the introduction of –NH$_2$ and –OH groups provided by EDTA and DMF [33, 34]. In addition, the intensity of the vibration peak at 1240 cm^{-1} of C–N (amide III) and C–H increased, and this peak broadened and shifted to 1259 cm^{-1}. This phenomenon was aroused by the introduced tertiary amine and a small amount of –COO$^-$ groups in DMF and EDTA. The signals of the stretching and vibration modes of C–O (1025 and 1155 cm^{-1}) also increased, which can be attributed as well to the added carboxyl groups from EDTA [35].

After these wings were immersed in AgNO$_3$ ethanol solutions, the intensity of the 1309 cm^{-1} absorption band of wing templates increased. Also, the symmetric deformation modes of –CH$_3$ at 1378 and 1450 cm^{-1} turned to a wider and stronger absorption band at 1383 cm^{-1} [36]. These phenomena were mainly caused by the polarity of AgNO$_3$ ethanol solutions that brought lots of ionized –COO$^-$. Moreover, the absorption band at 3280 cm^{-1} became broader. Considering that each EDTA

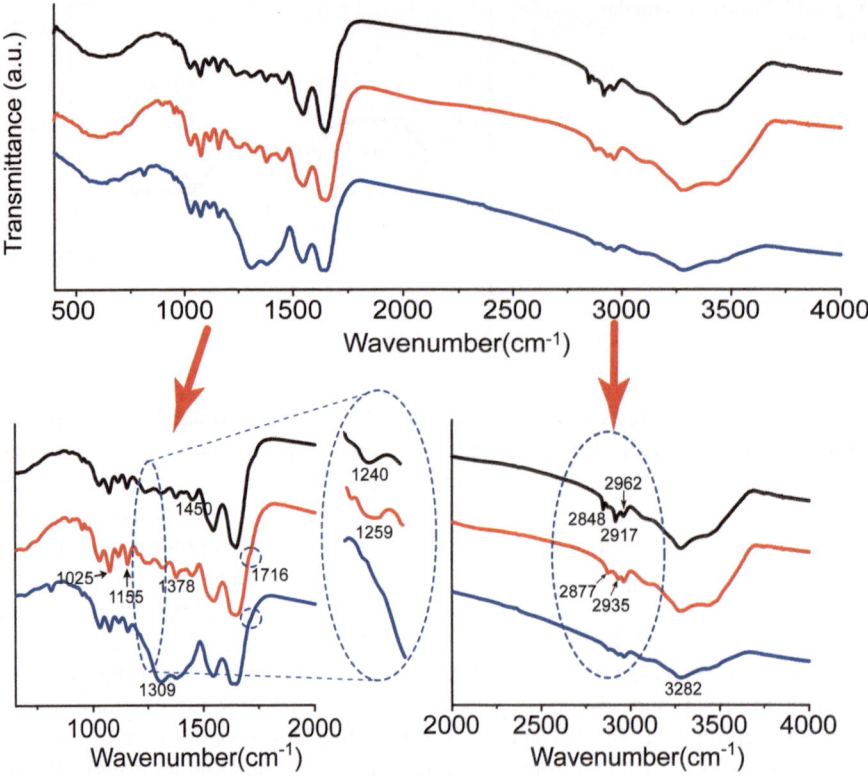

Fig. 2.14 FTIR spectra of original butterfly wings and the activated wings. (Reproduced with permission [32]. Copyright 2012, American Chemical Society)

molecule contains four ligands capable of chelating metal ions (including Ag^+) [37], this result indicates the formation of $-COO^-$, $-NH_3^+$, and in turn a large number of complexes [34].

Thus, the activation by EDTA/DMF can produce a large number of carboxyl groups which Ag^+ ions can complex with. This provides an important basis for the preparation of Ag replicas with a complete scale morphology.

2.4.3   Characterizations

After the pretreatment and photoreduction processes, Ag butterfly wings were obtained. Figure 2.15 shows the SEM observations on Ag replicas of *E. mulciber* and *P. paris*, respectively. Figure 2.15a and b are their secondary electron (SE) images. Main ridges and ribs on both samples were well-observed, which are the key photonic structures of an original butterfly wing. Moreover, the pore structures of *P. paris* scales were also preserved (Fig. 2.15b). Figure 2.15c and d provide the

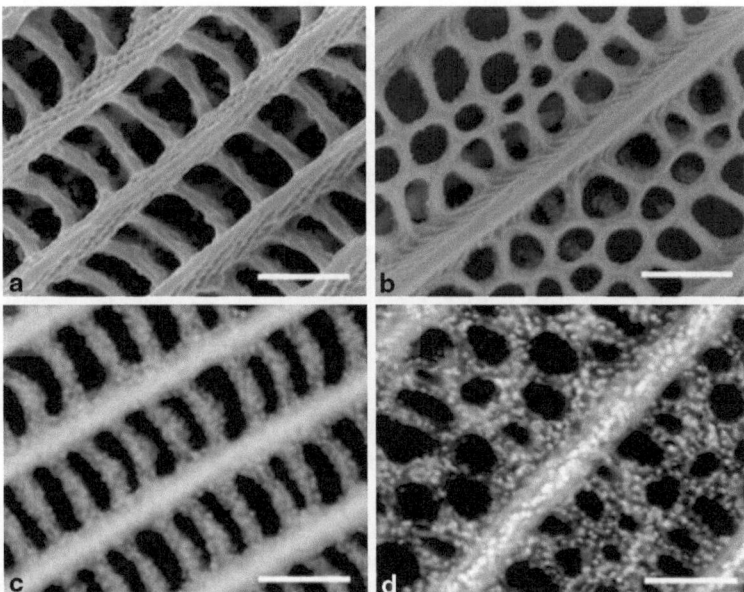

Fig. 2.15 Scanning electron microscopy (SEM) images of as-prepared Ag scale replicas. **a** and **b** show secondary electron (SE) images of the Ag replicas of (**a**) *E. mulciber* and (**b**) *P. paris*. **c** and **d** are backscattered electron (BSE) images of (**a**) and (**b**), respectively. Scale bars: 1 μm. (Reproduced with permission [32]. Copyright 2012, American Chemical Society)

backscattered electron (BSE) images of these two Ag replicas. The BSE images suggest that Ag layer with a certain thickness was formed on the ridges and rib structures (Fig. 2.15c) of *E. mulciber*. In addition, dense Ag NPs covered the hole rims of *P. paris* (Fig. 2.15d). Their small size ensured the distribution of Ag NPs across the scale surfaces, yielding a conformal replication of scale structures from nanoscales to macroscales.

This conclusion can be further proven in Fig. 2.16. We provide SEM images of the natural butterfly scales and their Ag replicas under comparatively low magnfications. Figure 2.16c and d confirm that the replication process can be fulfilled over a large area across the whole scale surface. Figure 2.17a and b are EDS and XRD results of these two Ag wing replicas. Four XRD peaks at 37.9°, 44.1°, 64.8°, and 77.5° were observed for the final products, which correspond to *fcc* Ag. No other phases were found in the XRD results. For *E. mulciber* and *P. paris*, both replicas shared the same NP size (~20 nm) according to the Scherrer equation. These results prove that this photoreduction route can generate Ag NPs with the same size on wings from various butterfly species.

More detailed structural features of Ag scales can be observed using TEM. Figure 2.18 provides the TEM images of Ag replicas of *E. mulciber*. As shown in Fig. 2.18a and b, Ag scales were composed of Ag NPs with a diameter of ~20 nm (Fig. 2.18c and d), which consists well with the XRD analyses. The SEAD pattern in Fig. 2.18a clearly shows a series of diffraction rings, which can be indexed as

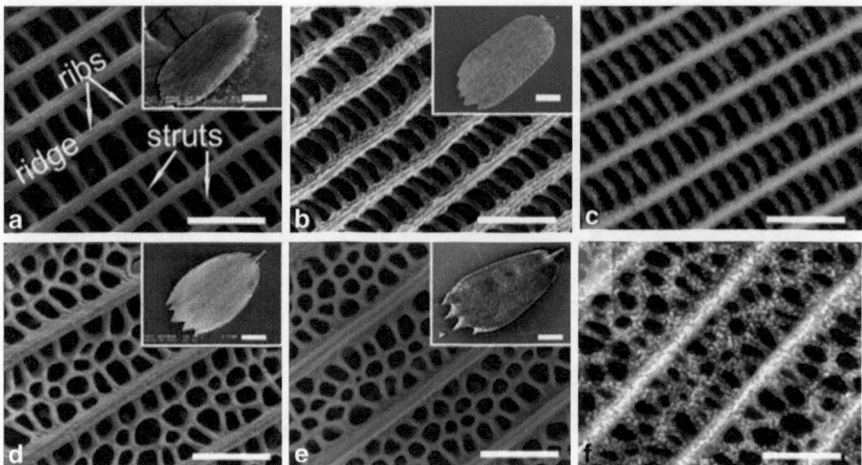

Fig. 2.16 Scanning electron microscopy (SEM) images of the natural butterfly wing scales and the corresponding Ag replicas. **a** and **b** are natural butterfly scales of (**a**) *E. mulciber* scale and (**b**) *P. paris*. **c** and **d** show SE images of (**a**) and (**b**), respectively. **e** and **f** provide corresponding backscattered electron (BSE) images of (**c**) and (**d**), respectively. Scale bars: 2 μm. Insets of (**a**)–(**d**) show the whole scales where corresponding SEM images were taken. Scale bars: 25 μm. (Reproduced with permission [32]. Copyright 2012, American Chemical Society)

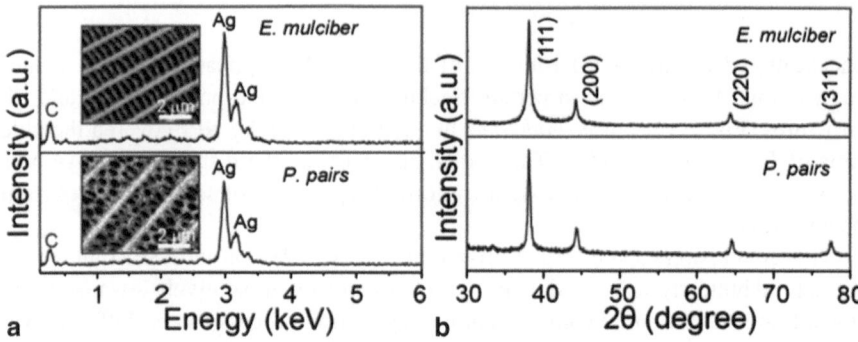

Fig. 2.17 **a** Energy-dispersive X-ray spectroscopy (EDS) results of Ag replicas of *E. mulciber* and *P. paris*. **b** Corresponding X-ray diffraction (XRD) results. (Reproduced with permission [32]. Copyright 2012, American Chemical Society)

(111), (200), (221), (331), and (222) plane of *fcc* Ag, respectively. Moreover, lattice fringes of an individual NP are shown in Fig. 2.18d. The lattice spacing is 0.24 nm, which corresponds to the (111) plane of *fcc* Ag.

Therefore, although the homogeneity in Ag NP distribution is yet to be improved (Fig. 2.15c and d), scale structures can basically be converted to Ag with the photoreduction method.

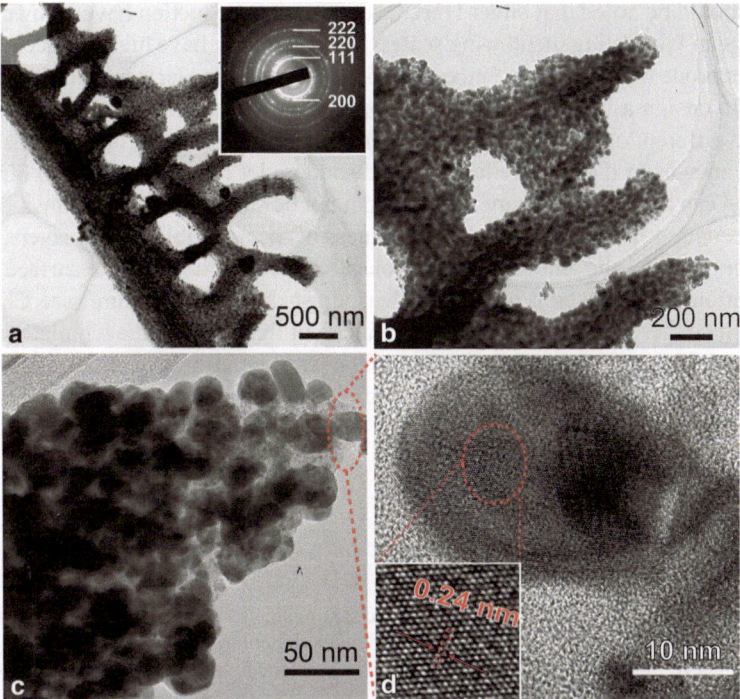

Fig. 2.18 Transmission electron microscopy (TEM) images of the Ag replica. Inset of (**a**) is a selected area electron diffraction (SAED) pattern. Inset of (**d**) demonstrates a Ag NP imaged along the <111> zone axis. The observed structures were shaken off from the Ag replicas and dispersed in ethanol via ultrasonic treatment. (Reproduced with permission [32]. Copyright 2012, American Chemical Society)

2.5 Summary

In this chapter, two kinds of butterfly wings with different typical structures were selected as biotemplates. Cu and Ag wing replicas were prepared via H_2 reduction and photoreduction, respectively. Main conclusions are the following:

To get Cu replicas, CuO scales (intermediates) with corresponding wing structures were first prepared via an impregnation plus calcination method. Then, the CuO films were sintered and reduced into Cu at 500 °C in a H_2 atmosphere. Although the reduced Cu films retained the morphologies of original scales to some extent, the finest scale structures could not be well-preserved.

To obtain Ag replicas, a direct photoreduction method conducted under *room temperature* was applied. Based on the specific wing components (e.g., chitin), the biotemplates were first activated (e.g., by EDTA/DMF) to coordinate more metal ions, which is beneficial to forming a continuous metal layer in the successive photoreduction process.

It should be noted that either H_2 reduction or photoreduction have its own disadvantage in fabricating metal wing scales. Metals usually have high surface energies and are prone to coarsening. Since various metals defer in electronegativity, it is difficult to design a versatile route with proper sintering temperatures to conformally reduce different oxide intermediates into metals. In comparison, there are also three disadvantages in the photoreduction conducted under room temperature. First, this method can be used to prepare Ag but might not be suitable for other transition metals. Second, it is hard to control the thickness of the assembled metal layers, which is determined by the amount of metal ions absorbed on the template surface before irradiation. This makes it hard to further adjust the replica texture (see Chap. 5). Last, since there are blind corners to the incident light cast on the template surface, the homogeneity in metal NP distribution is yet to be improved. Hence, a versatile route toward metal butterfly scales, which have well-controlled 3D morphologies down to a submicrometer level, is highly needed.

References

1. Zhu SM, Zhang D, Li ZQ et al (2008) Precision replication of hierarchical biological structures by metal oxides using a sonochemical method. Langmuir 24:6292–6299
2. Zhang W, Zhang D, Fan TX et al (2009) Novel photoanode structure templated from butterfly wing scales. Chem Mater 21:33–40
3. Zhang W, Zhang D, Fan TX et al (2006) Biomimetic zinc oxide replica with structural color using butterfly (*Ideopsis similis*) wings as templates. Bioinspir Biomim 1:89–95
4. Liu XY, Zhu SM, Zhang D et al (2010) Replication of butterfly wing in TiO_2 with ordered mesopores assembled inside for light harvesting. Mater Lett 64:2745–2747
5. Chen Y, Gu JJ, Zhang D et al (2011) Tunable three-dimensional ZrO_2 photonic crystals replicated from single butterfly wing scales. J Mater Chem 21:15237–15243
6. Wang J, Chen X, Wang G et al (2002) Melting behavior in ultrathin metallic nanowires. Phys Rev B 66:085408
7. Lisiecki I, Sack-Kongehl H, Weiss K et al (2000) Annealing process of anisotropic copper nanocrystals. Langmuir 16:8807–8808
8. Qin Y, Staedler T, Jiang X (2007) Preparation of aligned Cu nanowires by room-temperature reduction of CuO nanowires in electron cyclotron resonance hydrogen plasma. Nanotechnology 18:035608
9. Dippel M, Maier A, Gimple V et al (2001) Size-dependent melting of self-assembled indium nanostructures. Phys Rev Lett 87:095505
10. Link S, Wang ZL, El-Sayed MA (2000) How does a gold nanorod melt? J Phys Chem B 104:7867–7870
11. Payne EK, Rosi NL, Xue C et al (2005) Sacrificial biological templates for the formation of nanostructured metallic microshells. Angew Chem Int Ed 44:5064–5067
12. Garrett NL, Vukusic P, Ogrin F et al (2009) Spectroscopy on the wing: Naturally inspired SERS substrates for biochemical analysis. J Biophoton 2:157–166
13. Bao ZH, Ernst EM, Yoo S et al (2009) Syntheses of porous self-supporting metal-nanoparticle assemblies with 3D morphologies inherited from biosilica templates (diatom frustules). Adv Mater 21:474–478
14. Wang QQ, Han JB, Gong HM et al (2006) Linear and nonlinear optical properties of ag nanowire polarizing glass. Adv Funct Mater 16:2405–2408

15. Rivas L, Sanchez-Cortes S, García-Ramos JV et al (2000) Mixed silver/gold colloids: a study of their formation, morphology, and surface-enhanced Raman activity. Langmuir 16:9722–9728
16. Sun YG, Lei CH (2009) Synthesis of out-of-substrate Au-Ag nanoplates with enhanced stability for catalysis. Angew Chem Int Ed 48:6824–6827
17. Wang W, Yang Q, Fan F et al (2011) Light propagation in curved silver nanowire plasmonic waveguides. Nano Lett 11:1603–1608
18. Drachev VP, Nashine VC, Thoreson MD et al (2005) Adaptive silver films for detection of antibody-antigen binding. Langmuir 21:8368–8373
19. Bao ZH, Weatherspoon MR, Shian S et al (2007) Chemical reduction of three-dimensional silica micro-assemblies into microporous silicon replicas. Nature 446:172–175
20. Kinoshita S, Yoshioka S, Miyazaki J (2008) Physics of structural colors. Rep Prog Phys 71:076401
21. Mason CW (1925) Structural colors in insects. I. J Phys Chem 30:383–395
22. Mason CW (1926) Structural colors in insects. II. J Phys Chem 31:321–354
23. Chen Y, Gu JJ, Zhu SM et al (2009) Iridescent large-area ZrO_2 photonic crystals using butterfly as templates. Appl Phy Lett 94:053901
24. Tan YW, Gu JJ, Xu W et al (2013) Reduction of cuo butterfly wing scales generates Cu SERS substrates for DNA base detection. ACS Appl Mater Interfaces 5:9878–9882
25. Vukusic P, Sambles JR, Lawrence CR (2004) Structurally assisted blackness in butterfly scales. Proc Biol Sci Roy Soc 271 Suppl 4:4S237–S239
26. Koon DW, Crawford AB (2000) Insect thin films as sun blocks, not solar collectors. Appl Opt 39:2496–2498
27. Heilman BD, Miaoulis LN (1994) Insect thin films as solar collectors. Appl Opt 33:6642–6647
28. Han J, Su HL, Zhang D et al (2009) Butterfly wings as natural photonic crystal scaffolds for controllable assembly of cds nanoparticles. J Mater Chem 19:8741–8746
29. Richards AG (1947) Studies on arthropod cuticle. I. The distribution of chitin in lepidopterous scales, and its bearing on the interpretation of arthropod cuticle. Ann Entomol Soc Am 40:227–240
30. Ravi Kumar MNV (2000) A review of chitin and chitosan applications. Reactive Funct Polym 46:1–27
31. Mathew AP, Laborie M-PG, Oksman K (2009) Cross-linked chitosan/chitin crystal nanocomposites with improved permeation selectivity and pH stability. Biomacromolecules 10:1627–1632
32. Tan YW, Zang XN, Gu JJ et al (2011) Morphological effects on surface-enhanced Raman scattering from silver butterfly wing scales synthesized via photoreduction. Langmuir 27:11742–11746
33. Lanigan KC, Pidsosny K (2007) Reflectance FTIR spectroscopic analysis of metal complexation to EDTA and EDDS. Vib Spectrosc 45:2–9
34. Bellamy LJ (1980) The infrared spectra of complex molecules. Chapman and Hall Ltd., London
35. Langer HG (1963) Infrared spectra of ethylenediaminetetraacetic acid (EDTA). Inorg Chem 2:1080–1082
36. Satroutdinov AD, Dedyukhina EG, Chistyakova TYI et al (2000) Degradation of metal-EDTA complexes by resting cells of the bacterial strain DSM 9103. Environ Sci Technol 34:1715–1720
37. Chen L, Liu T, Ma CA (2009) Metal complexation and biodegradation of EDTA and S, S-EDDS: A density functional theory study. J Phys Chem A 114:443–454

Chapter 3
Metal Scale Replicas Prepared via Electroless Deposition

3.1 Introduction

Direct deposition methods include physical vapor deposition (PVD), chemical vapor deposition (CVD), and electroplating or electroless plating, etc. [1, 2]. PVD describes a variety of vacuum deposition methods used to deposit thin films onto a workpiece surface by the condensation of a vaporized material. These methods (e.g., magnetron sputtering and plasma bombardment) involve purely physical processes. In comparison, CVD is based on chemical reactions happening at the surface to be coated. However, the line-of-sight nature of both PVD and CVD prevents a complete replication of original 3D biomorphologies [3]. This shading effect can be avoided by metal plating conducted in liquid environments. Electroplating uses electrical current to reduce dissolved metal cations and can form a coherent metal coating on a conductive substrate. However, since biotemplates are usually nonconductive, it is unsuitable to the replication of biostructures.

Thus, electroless plating is applied to fabricate metal scale replicas. As an autocatalytic chemical technique, electroless plating involves reducing agents such as sodium citrate, sodium borohydride ($NaBH_4$), glucose, tartaric acid, and ascorbic acid, etc. [4–7], which may reduce the metal ions from plating solutions. The resulting metal atoms will nuclear and grow up to form nanoparticles (NPs) with different sizes and morphologies. Generally, no external electric currents are required during the plating process. Coatings produced by electroless plating are uniform, continuous, and tunable, which makes this process very attractive in industry. Compared with electroplating, electroless deposition has several advantages. First, the process itself is simple and compact. Second, it can produce homogeneous coatings across the whole surface. Electroplating might cause excessive build-ups of materials on the edges and corners of a workpiece, while electroless plating can produce uniform coatings that evenly cover the entire surface with a complex texture. Third, the coatings generated by electroless plating are compact and less porous, which is helpful for the formation of free-standing metal replicas. Finally, depositing metals onto nonconductive organic templates can be realized by this method. Up to now, electroless plating has been widely applied, especially in the deposition of functional

© Jiajun Gu, Di Zhang, and Yongwen Tan 2015
J. Gu et al., *Metallic Butterfly Wing Scales,* SpringerBriefs in Materials,
DOI 10.1007/978-3-319-12535-0_3

coatings on organic nanostructures with complex textures. Since the main component of butterfly wing scales is chitin, which is the second richest natural macromolecular compounds [8], it is feasible to deposit metal NPs onto butterfly wing scales via electroless plating.

In this chapter, we will demonstrate a versatile method to prepare metal scale replicas via electroless plating. This low-temperature approach comprises a surface functionalization step, an electroless deposition step, and a template removal step. It can replicate chitin-based scales in at least seven important metals (i.e., cobalt, nickel, copper, palladium, silver, platinum, and gold). The influence exerted by the synthesis conditions on the obtained metal structures will also be discussed in details.

3.2   Sample Preparation

Figure 3.1 illustrates the fabrication process of metal butterfly wing scales with submicrometer structures via electroless plating. The scales of *Euploea mulciber* were used as biotemplates and their structures have been described in the last chapter.

To prepare the metal replicas, the wings of *E. mulciber* were first immersed in a dilute (8 vol%) HNO_3 water solution for 2 h, washed in deionized water, and dried in air. Then, the wings were immersed into an ethanol solution of ethanediamine (25 vol%) for 6 h (amination) and subsequently washed with ethanol. The aminated samples were put into an aqueous solution of $HAuCl_4$ (0.2 wt%) for 4 h followed by a rinse in deionized water. Au-NP seeds as catalysts were thus formed on the aminated biosurface. The samples were then exposed in 0.1 M aqueous solution of $NaBH_4$ for 120 s to reduce the bound Au^{3+} ions, washed with deionized water, and dried in air.

To deposit different kinds of metals, the Au-NP-functionalized butterfly wings were immersed in corresponding electroless plating solutions at modest temperatures, rinsed with deionized water, and dried in air. Detailed plating solutions for various metals and corresponding reaction conditions are listed in Table 3.1. During the electroless deposition process, there were observable tiny H_2 bubbles generated for all the metals except Ag and Au. These bubbles were removed from the sample surface by carefully touching the sample edges with a glass rod, making the morphology inconsistence originating from the H_2 bubbles negligible. Single scales with metal NPs deposited were subsequently removed from the as-plated wings onto a silicon wafer using a steel needle tip, and each scale was manipulated and arranged with its basal surface attached on Si wafers by surface tensions. This transfer process was conducted under an optical stereomicroscope. Finally, an aqueous solution of phosphoric acid (85 wt%) was dripped onto the metal-NP-coated scales with a needle tube. The samples were then kept in a sealed vessel at room temperature for 72 h to remove the original biotemplates. After that, samples were carefully washed in deionized water and the 3D metal replicas of original scales were obtained.

The thickness of metal layers can be controlled by regulating experimental parameters, such as the deposition time, temperature, pH value, and the concentration

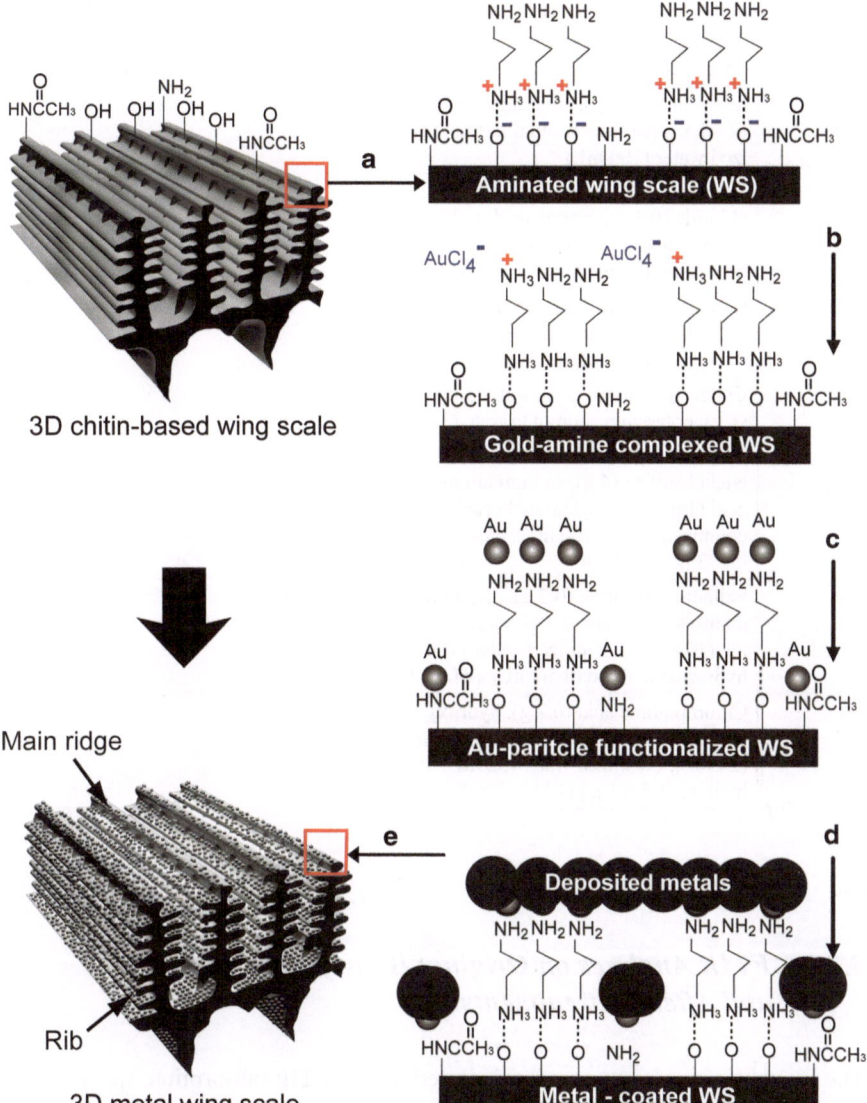

Fig. 3.1 Conformal synthesis route for metal scales. The original scale was subsequently treated with **a** EDA, **b** HAuCl$_4$, **c** NaBH$_4$, **d** electroless deposition solutions, and **e** H$_3$PO$_4$. Schematics of original scale structure and resulting metal replica are shown on the left side. (Reproduced with permission [9] Copyright 2011, Wiley-VCH)

of plating solutions. Such adjustments will change the morphology of the as-synthesized replicas and can exert effects on their related performance, which will be discussed in details in Chap. 5.

Table 3.1 Detailed compositions of electroless deposition solutions and corresponding reaction conditions. (Reproduced with permission [9] Copyright 2011, Wiley-VCH)

Materials	Solutions	Time (min)	Temperature (°C)
Ag	Silver nitrate (1 g), ammonium hydroxide (2 mL), potassium sodium tartrate (5 g), deionized water (100 mL)	20	25
Au	Hydrogen tetrachloroaurate (0.5 g), sodium chloride (0.3 g), tartaric acid (0.20 g), sodium hydroxide (2.57 g), ethanol (3.5 mL), deionized water (50 mL)	15	25
Co	Cobalt Sulfate (2.5 g), sodium succinate (2.5 g), dimethylamine borane (0.4 g), deionized water (100 mL)	15	25
Cu	Copper sulfate (0.75 g), sodium hydroxide (1 g), potassium sodium tartrate (3.5 g), deionized water (25 mL), formaldehyde (2.5 mL)	10	25
Ni	Nickel sulfate (4 g), sodium citrate (2 g), lactic acid (1 g), dimethylamine borane (0.2 g), deionized water (100 mL), ammonium hydroxide (3 mL); pH 6.8	20	25
Pd	Palladium chloride (0.2 g), disodium ethylenediamintetraacetate (4.1 g), ammonium hydroxide (19.8 mL), deionized water (100 mL), hydrazine hydrate (0.56 mL), pH 10.2	20	25
Pt	Chloroplatinic acid (0.2 g), hydroxy lamine hydrochloride (0.32 g), ammonium hydroxide (24 mL), deionized water (52 mL), hydrazine hydrate (1.5 mL), pH 11	300	50

3.3 Replication Mechanisms

3.3.1 FTIR Analyses on Original Butterfly Wings Before and After Pretreatments

The synthesis processes were monitored using FTIR absorption spectra. The changes in functional groups for the original butterfly wings, ethanediamine (EDA) aminated scales, and $HAuCl_4$ treated (Au-amine complexed) scales are provided in Fig. 3.2. As introduced above, –OH, –NHCOCH$_3$, and –NH$_2$ groups are naturally exposed on scale surfaces but not abundant enough to effectively coordinate metal ions. To achieve uniform and complete metal coatings at a submicrometer level on the hierarchically-structured wing surface, one can either expose more such groups by pretreating the templates [10, 11], or directly link additional groups like –NH$_2$ to the surfaces. These additional groups are supposed to complex with more metal ions. For example, EDA containing two amino groups can be applied to aminate the scale surface first. One –NH$_2$ group of EDA may bind with a

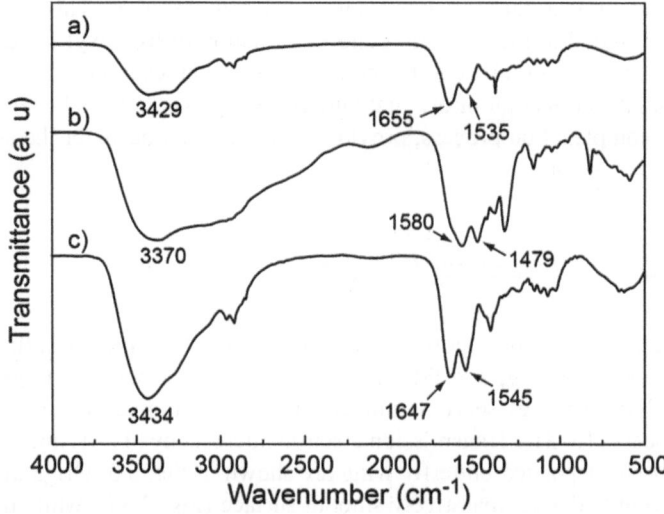

Fig. 3.2 FTIR spectra of **a** original butterfly wing scales, **b** EDA aminated scales, and **c** HAuCl₄ treated (Au-amine complexed) scales, respectively. (Reproduced with permission [9] Copyright 2011, Wiley-VCH)

–OH group on the chitin-based surface by nucleophilic attack, leaving the other in complex with Au^{3+}. Au NPs are subsequently formed at the aminated scale surface via the reduction of the coordinated Au^{3+} by $NaBH_4$. These Au NPs could act as catalysts for the subsequent electroless metal deposition. A detailed analysis of the functionalization process based on FTIR spectra are shown in Fig. 3.2. The characteristic wide band at 3429 cm^{-1} corresponds to the stretching vibration modes of –NH_2, –NH–, and –OH groups [12], while the peaks at 1655 and 1535 cm^{-1} are attributed to C=O stretching (amide I band) and N–H bending (amide II band) [13], respectively. When ethylenediamine was added, the peak at 3429 cm^{-1} shifted to lower wave number at 3370 cm^{-1} and became wider and stronger [14]. This result suggests that there were more associated bodies due to the hydrogen bonds formed between amino groups and hydroxyl groups from chitin. Since amine salt's characteristic FTIR feature is a wide absorption band with one or more peaks around 2500–3000 cm^{-1} [15], protonized amino groups might originate from the alkali ionization of ethylenediamine.

The peak at 3370 cm^{-1} shifted back to 3434 cm^{-1} and became sharper after HAuCl₄ was introduced. The amino groups were in complex with Au^{3+}, reducing the associated bodies and hydrogen bonds. Meanwhile, the added ethylenediamine increased the primary amine and even secondary amine (ethylenediamine alkylation). Thus, the amine stretching peak near 3370 cm^{-1} was strengthened as compared to the amide I and II bands of original butterfly wings. This enhancement may even exist after the amine-Au complexation process.

As additional supporting evidence, after the EDA was introduced, the peaks corresponding to amide I at 1655 cm^{-1} and amide II at 1535 cm^{-1} shifted to lower wave numbers and moved back after the amine-Au complexation. This phenomenon could be attributed to the formation and deformation of hydroxyl bonds between amino and carbonyl groups due to the amine-Au complexation. All these results indicate that the amination of original butterfly wings via EDA is beneficial to the successive complexation process, providing an important basis for the subsequent electroless deposition.

3.3.2 Surface Treatment Using Au Nanoparticles

Plating metal particles on a polymer surface is difficult. Typically, a thin layer of metal NPs (e.g., Au, Ag, and Pd) can be first deposited onto the target surface. These catalytic particles serve as nucleation sites for the subsequent electroless plating process [16–21]. Transmission electron microscopy (TEM) images of original and Au-NP-deposited butterfly wing are shown in Fig. 3.3. Original chitinous butterfly wing had a comparatively smooth surface (Fig. 3.3d), while the surface was covered by a layer of uniform Au NPs ($\sim$4 nm in size, Fig. 3.4) after the amination and Au activation processes. In energy-dispersive X-ray spectroscopy (EDS) analyses, only C, O, Au, and Cu signals from TEM grids can be clearly detected

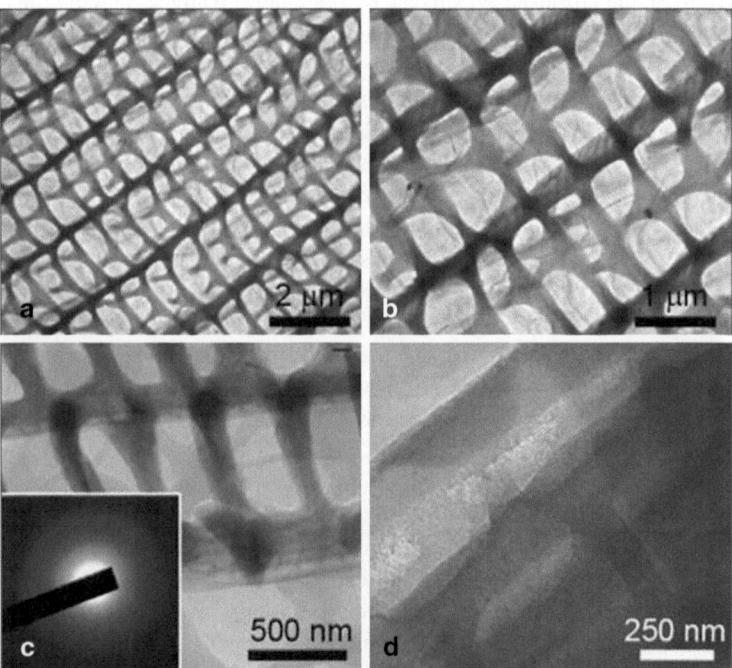

Fig. 3.3 Transmission electron microscopy (TEM) images of an original butterfly scale. Inset of (**c**) shows a selected area electron diffraction (SAED) pattern of the sample

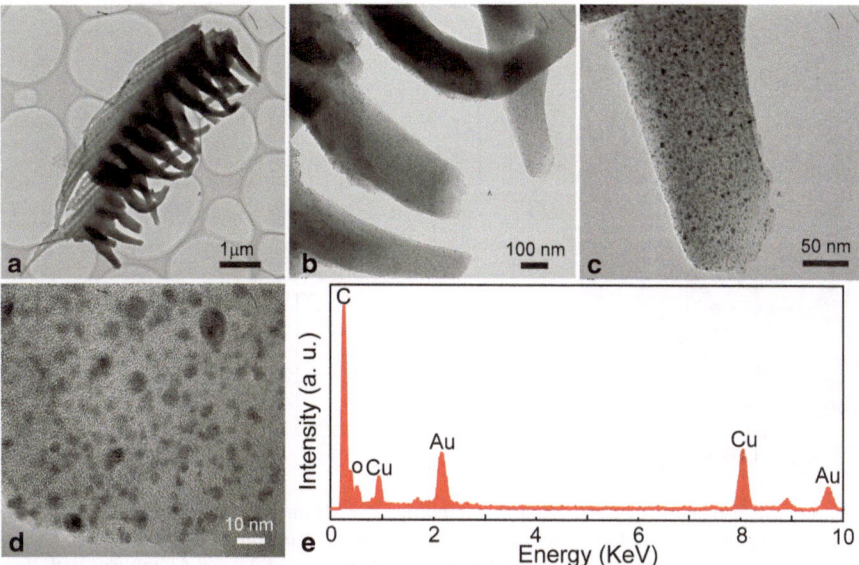

Fig. 3.4 a–d Transmission electron microscopy (TEM) images of butterfly wings activated by Au NPs. **e** Energy-dispersive X-ray spectroscopy (EDS) spectrum

Fig. 3.5 X-ray diffraction (XRD) results of original (*bottom spectrum*) and Au-NP-deposited (*top spectrum*) butterfly wings. Both spectra showed the same chitin peak located at 24.8°

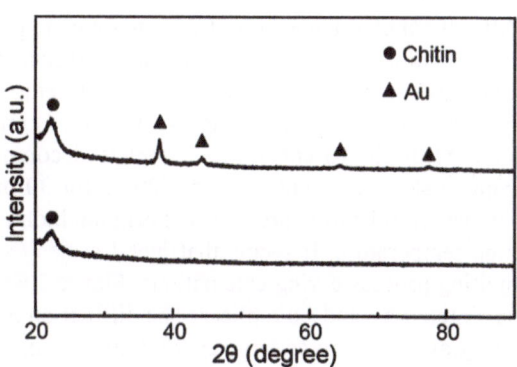

on the Au-NP-deposited butterfly wing (Fig. 3.4e). Such results are consistent with the X-ray diffraction (XRD) analyses provided in Fig. 3.5. For Au-NP-deposited samples, there are three main diffraction peaks at 38.2, 44.4, and 64.6°, which correspond to (111), (200), and (220) plane of Au, respectively [22]. According to Scherrer equation, the average grain size of Au NPs is ~4.2 nm, which agrees well with the Au crystallites observed in Fig. 3.4.

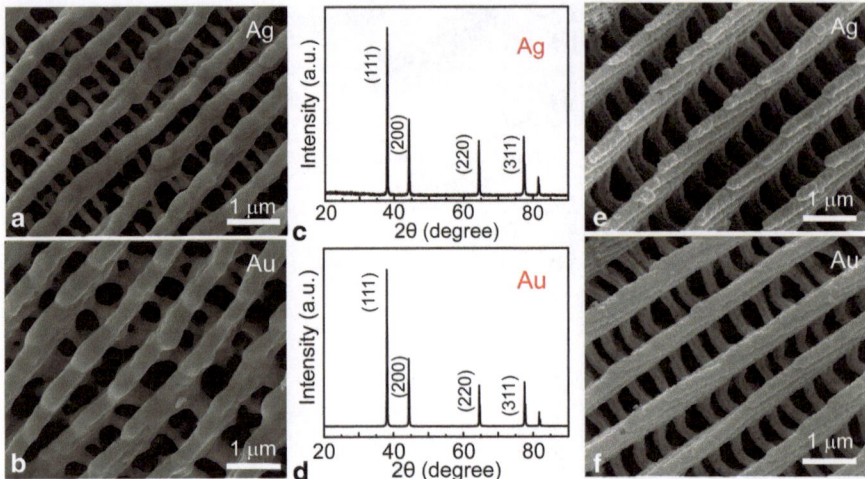

Fig. 3.6 **a** and **b** show scanning electron microscopy (SEM) images of Ag and Au replicas with their biotemplates removed via calcinations. **c** and **d** are corresponding X-ray diffraction (XRD) results of (**a**) and (**b**), respectively. **e** and **f** provide SEM images of Ag and Au replicas with their biotemplates removed via acid treatment under mild temperatures

3.3.3 Removal of Biotemplates

After the metal deposition, it is sometimes important to remove the biotemplates to get complete metal replicas. Usually, original bioskeletons could be removed by calcinations in muffle furnace under high temperatures [23–27]. Figure 3.6a and b show scanning electron microscopy (SEM) images of Ag and Au scale replicas sintered in air. To ensure the complete decomposition of organic templates, the wing scales were kept in air at 500 °C for 30 min. Unfortunately, these delicate replicas could barely preserve the original biomorphology after experiencing such high temperature. It seems that metal scales experienced severe coarsening and melting process during calcinations. Figure 3.6c and d are the XRD results of the as-sintered Ag and Au replicas. The diffraction peaks are rather sharp and the average grain sizes are as large as 70.2 and 62 nm (Scherrer equation), respectively. This indicates as well that metal NPs are prone to aggregate under the high temperature [28–31]. Thus, it is almost impossible to keep the original biostructures after the heat treatment, let alone the fact that some metals are easily oxidized during calcinations.

In comparison, it is a better choice to chemically melt biotemplates off. An aquarous solution of H_3PO_4, for example, can used to remove organic chitinous templates at low temperature [32, 33]. Figure 3.6e and f show SEM images of Ag and Au replicas, respectively, with their biotemplates dissolved in H_3PO_4. Important wing scale structures such as main ridges and ribs were well-preserved. Therefore, this low-temperature approach can help produce complete metal replicas without losing structural features.

Fig. 3.7 Thermal gravimetric analysis (TGA) on biological wing scales, chitin-based scales covered with noble metal NPs, and noble metal replicas exposed to H_3PO_4, respectively

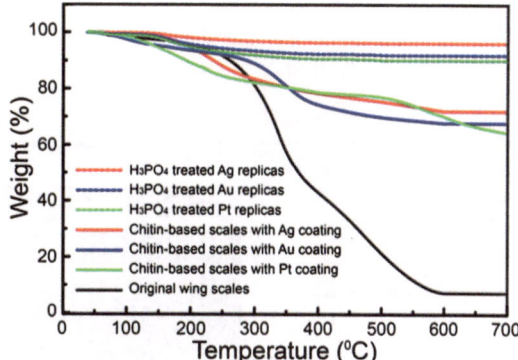

To determine the amount of biological residues after the H_3PO_4 treatment, thermal gravimetric analysis (TGA) was conducted on Ag, Pt, Au-coated samples before and after the H_3PO_4 treatment. An original wing was also selected as a control sample. As shown in Fig. 3.7, Ag, Pt, and Au coated samples kept 75, 70, and 67 % of their original weight after calcination, respectively, while the final yields of Ag, Pt, and Au replicas in weight were 96, 93, and 90 %, respectively, after the removal of biotemplates. These results indicate that most of the biostructures were successfully removed. The different yield of metals may result from the deferring thickness of metal layers on biotemplates.

3.4   Characterizations on as-Prepared Metal Replicas

3.4.1   Morphologies

Figure 3.8 provides SEM element mapping results of seven synthesized metal scales manually arranged together on a Si wafer. The scale color based on the mapping results was directly generated by a software (Microanalysis Suite, Oxford Instruments Inca) equipped on the SEM. All of the metal replicas were studied under high vacuum without Au sputtering, which was required for original non-conductive butterfly scales. Signals of Au NPs as catalysts could not be found in the replicas (except for the Au scale) as they were covered by the continuous metal coatings.

To evaluate the conformity between original butterfly scales and the final metal replicas, the metal scale morphologies are compared with their original counterparts in Fig. 3.9. Like many other butterflies and moths, the original scale of *E. mulciber* with hierarchical morphology usually has three levels of periods. The largest period (~700 nm) is defined by the spacing between two adjacent main ridges (Fig. 3.9a), while a smaller one (~380 nm) is the distance between struts (top view). Much finer structures (i.e., ribs) can be identified on individual ridges, with a period of

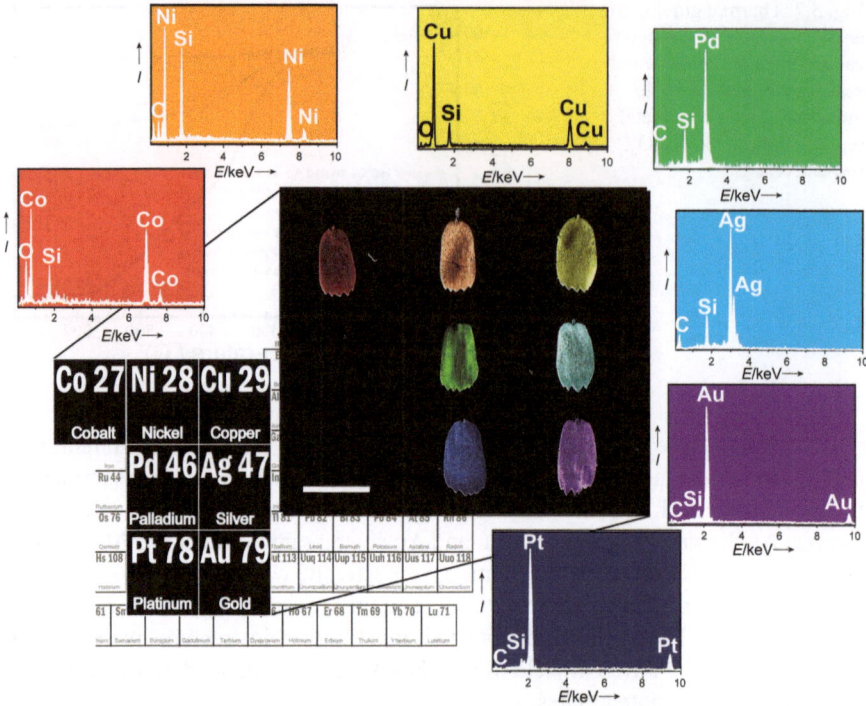

Fig. 3.8 Scanning electron microscopy (SEM) element mapping images of seven as-fabricated metal scale replicas. Si signals originated from the Si wafer. Scale bar: 50 μm. (Reproduced with permission [9] Copyright 2011, Wiley-VCH)

~ 100 nm in spacing. Due to the intrinsic structural differences in individual single scales, these values could slightly vary. Through this wet-chemical synthesis route, all these periodicities were topologically copied (Fig. 3.9b–h) for all the metal replicas owing to the capability of the metal ions to get into the blind corners of the intricate biostructures. It should be noted that the gap distances shown in Fig. 3.9b–h somewhat decreased because of the thickness of metal coatings. For example, the metal layers on main ridges and struts are ~ 20–50 nm in thickness, resulting in a shrinkage of ~ 6–14 and ~ 11–26 % in size for the ridge spacing and strut spacing, respectively. These dimensions could be further tuned by modifying the deposition conditions, which will in turn affect the replica performance and will be discussed in Chap. 5. By tilting the SEM stage for 45° against the length direction of main ridges, the coating on ribs is roughly estimated to be 20–30 nm in thickness and may cause a 40–60 % shrinkage in rib spacing. These results confirm that electroless plating can conformally copy wing scale structures from nanoscales to macroscales. Moreover, low-magnification SEM images (× 10,000, see Fig. 3.10) reveal a homogeneous metal deposition over a large surface area.

The metal scales were further manually fractured with a needle to observe their inner structures (Fig. 3.11). As described by many previous studies [34], the scale

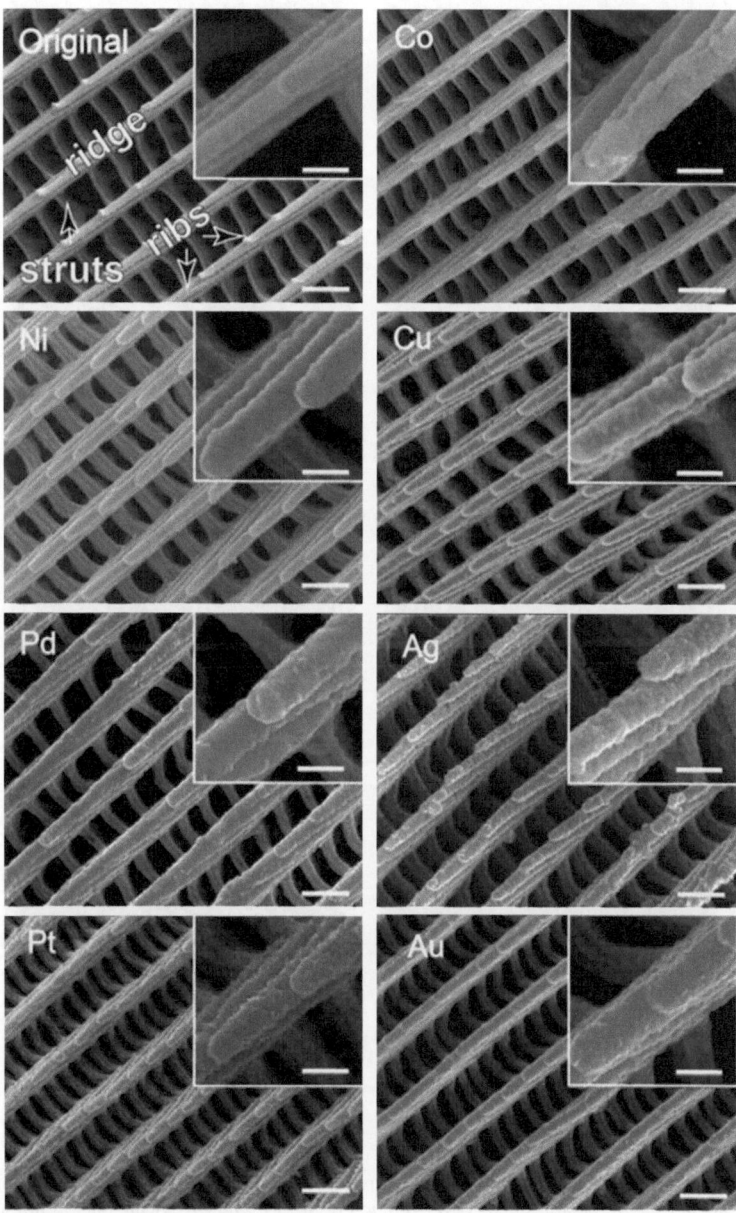

Fig. 3.9 High-magnification scanning electron microscopy (SEM) images of an original butterfly wing scale and seven metal replicas (*Co, Ni, Cu, Pd, Ag, Pt, and Au*). Zoomed-up observations are provided in corresponding insets. Scale bars: 5 μm. 250 nm (*insets*). (Reproduced with permission [9] Copyright 2011, Wiley-VCH)

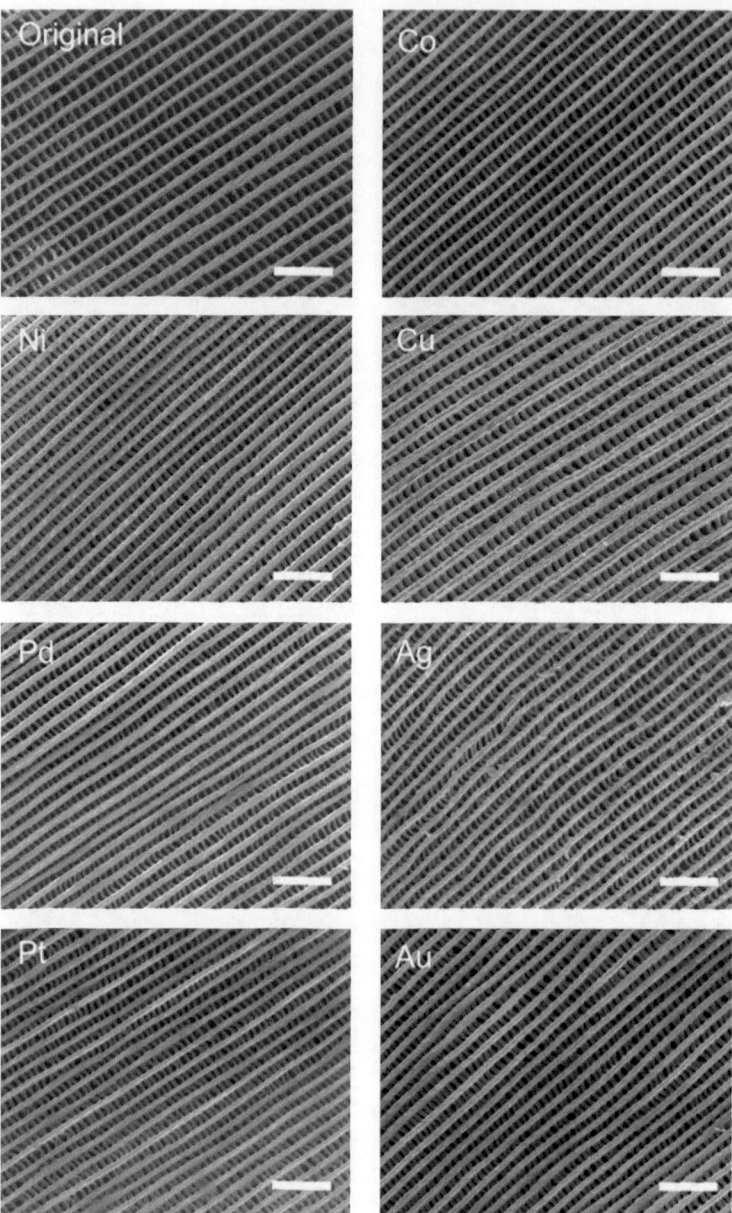

Fig. 3.10 Low-magnification scanning electron microscopy (SEM) images of an original butterfly wing scale and seven metal replicas (*Co, Ni, Cu, Pd, Ag, Pt, and Au*). Scale bars: 5 μm. (Reproduced with permission [9] Copyright 2011, Wiley-VCH)

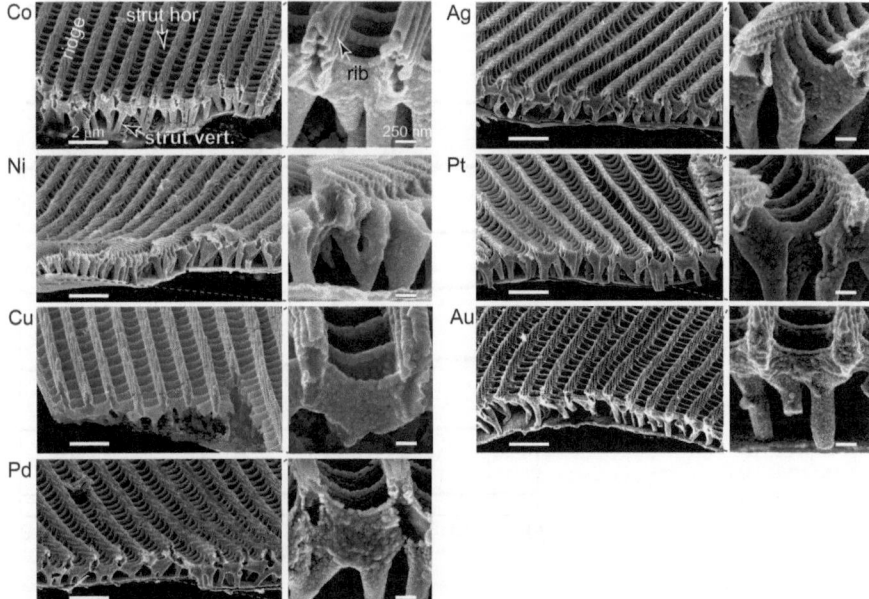

Fig. 3.11 Scanning electron microscopy (SEM) images (*side-view*) of seven metal replicas (*Co, Ni, Cu, Pd, Ag, Pt, and Au*). Scale main ridges, struts (*vertical and horizontal*), and ribs are denoted. (Reproduced with permission [9] Copyright 2011, Wiley, VCH)

struts actually comprise vertical and horizontal parts, which support and connect the main ridges. A close-up view of the structures (right column, Fig. 3.11) indicates that the replicas are hollow, owing to the removal of the original bioskeletons using H_3PO_4.

3.4.2 Phases

Figure 3.12 shows XRD results of Ag, Au, Co, Cu, Ni, Pd, and Pt replicas. The average grain size for individual sample can be calculated using Scherrer equation (Table 3.2). The size ranges from ~4 nm for Ni to ~20 nm for Ag, depending on the different metals and fabrication parameters. The small size of metal NPs promised the successful replication of wing scales' subtle morphologies down to the submicrometer level.

Fig. 3.12 X-ray diffraction
(XRD) spectra of seven metal
replicas. (Reproduced with
permission [9] Copyright
2011, Wiley, VCH)

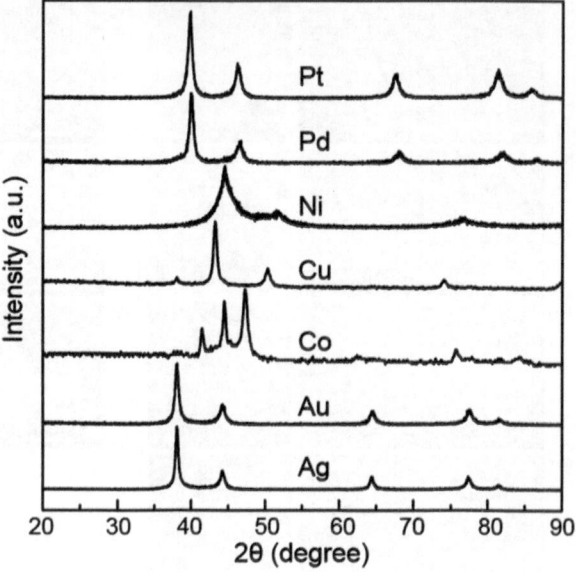

Table 3.2 X-ray diffraction (XRD) analyses on Ag,
Au, Co, Cu, Ni, Pd, and Pt
replicas

Metal	Grain size (nm)	Plane indexes	Type
Ag	20.4	(111) (200) (220) (311) (222)	*fcc*
Au	13.8	(111) (200) (220) (311) (222)	*fcc*
Co	12.7	(110) (002) (101) (102) (110)	*hcp*
Cu	17.1	(111) (200) (220)	*fcc*
Ni	3.80	(111) (200) (220)	*fcc*
Pd	12.5	(111) (200) (220)	*fcc*
Pt	13.1	(111) (200) (220) (311) (222)	*fcc*

Figure 3.13 provides TEM images of the Cu, Pd, Pt, and Au scale replicas.
The replica surfaces were uniformly covered by a layer of metal NPs. These NPs
preserved the hierarchical 3D morphologies of original wings, either for the submicrometer main ridges or for the nanosized ribs. In addition, the TEM observations
and the diffraction patterns verified the components and the crystallization form of
samples.

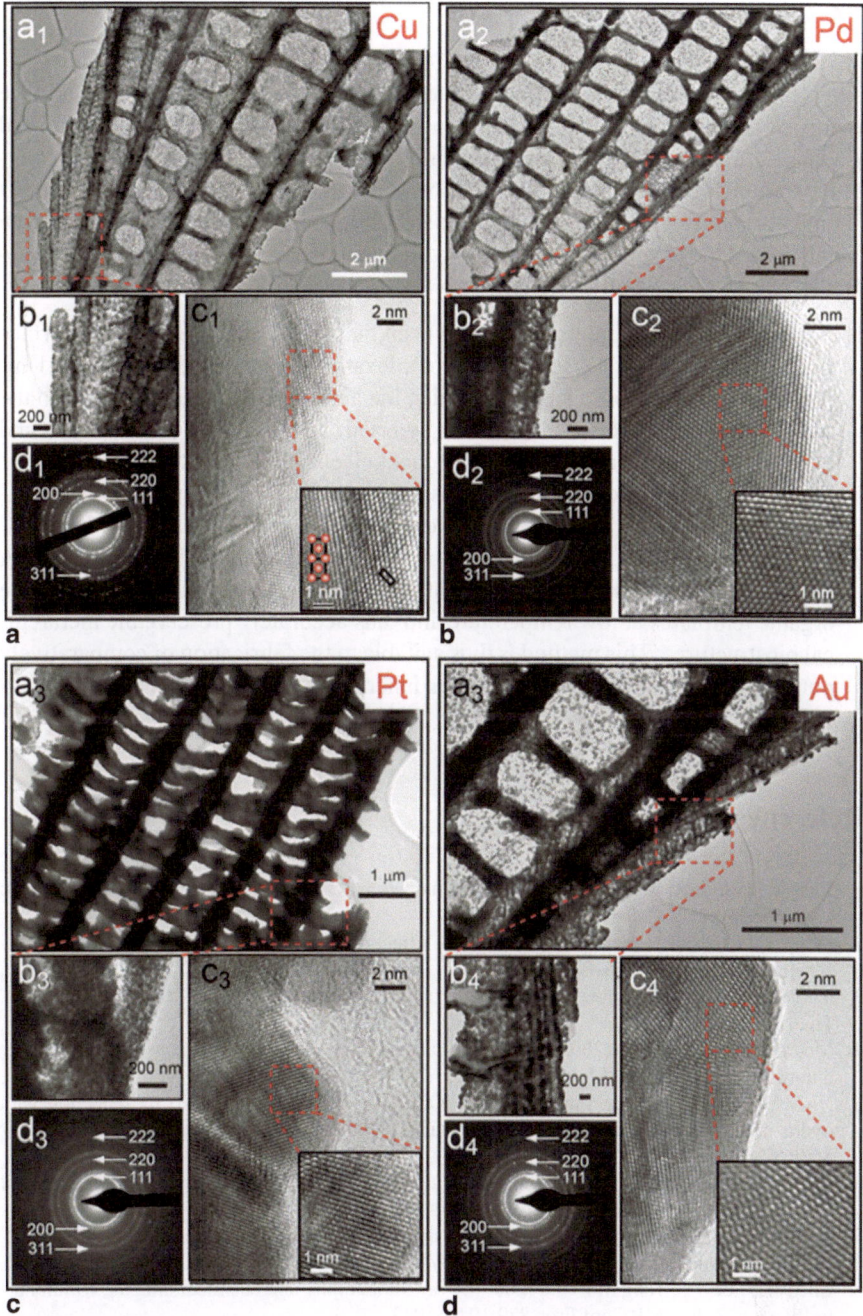

Fig. 3.13 a–c transmission electron microscopy (TEM) images of various metal replicas. Inset of each panel (**c**) shows lattice fringes of metal replicas. Each panel (**d**) provides selected area electron diffraction (SAED) patterns of corresponding samples

3.5   Summary

Since the 3D structures of original butterfly scales are complex and the surface energy of metals are high, the H_2 reduction and photoreduction methods introduced in Chap. 2 are not general enough to obtain a broad range of metal replicas. A versatile approach based on electroless deposition is thus developed to prepare intact 3D metal replicas of butterfly scales. This strategy has succeeded at least in seven metals (Ag, Au, Co, Cu, Ni, Pd, and Pt) with 3D photonic structures. The whole approach can be divided into three steps:

1. Surface activation: Agents (e.g., EDA) can be used to aminate the wing surface first. This will benefit a successive complexation process with Au^{3+} ions. Uniform Au NPs are subsequently formed at the aminated scale surface through the reduction of the coordinated Au^{3+} ions by $NaBH_4$. These NPs act as catalysts and provide substantial basis for the subsequent electroless deposition.
2. Standard electroless deposition: In this step, uniform and continuous metal coatings can be achieved down to a submicrometer level.
3. Biotemplate removal: This is realized by H_3PO_4 treatment conducted under room temperature. Compared with calcinations, chemical dissolution can remove organic templates at low temperature and thus accurately preserve the hierarchical biostructures. This method is thus suitable to the fabrication of comparatively pure metal replicas with fine structures inherited from butterfly scales.

References

1. Jinshan H, Solanki R, Mcandrew J (2002) Characteristics of copper films produced via atomic layer deposition. J Mater Res 17:2394–2398
2. Duffy J, Pearson L, Paunovic M (1983) The effect of pH on electroless copper deposition. J Electrochem Soc 130:876–880
3. Rossnagel SM (1995) Directional and preferential sputtering-based physical vapor deposition. Th Sol Films 263:1–12
4. Métraux GS, Mirkin CA (2005) Rapid thermal synthesis of silver nanoprisms with chemically tailorable thickness. Adv Mater 17:412–415
5. Nickel U, Zu Castell A, Pöppl K et al (2000) A silver colloid produced by reduction with hydrazine as support for highly sensitive surface-enhanced Raman spectroscopy. Langmuir 16:9087–9091
6. Gu X, Nie C, Lai Y et al (2006) Synthesis of silver nanorods and nanowires by tartrate-reduced route in aqueous solutions. Mater Chem Phys 96:217–222
7. Adhikari B, Banerjee A (2010) Facile synthesis of water-soluble fluorescent silver nanoclusters and Hg-II sensing. Chem Mater 22:4364–4371
8. Chen JH, Lee YC, Tang MT et al (2007) X-ray tomography and chemical imaging within butterfly wing scales. AIP Conf Proc 879:1940–1943
9. Tan Y, Gu J, Zang X et al (2011) Versatile fabrication of intact three-dimensional metallic butterfly wing scales with hierarchical sub-micrometer structures. Angew Chem Int Ed Engl 50:8307–8311

10. Zhang W, Zhang D, Fan TX et al (2009) Novel photoanode structure templated from butterfly wing scales. Chem Mater 21:33–40
11. Han J, Su HL, Zhang D et al (2009) Butterfly wings as natural photonic crystal scaffolds for controllable assembly of cds nanoparticles. J Mater Chem 19:8741–8746
12. Anandhavelu S, Thambidurai S (2011) Preparation of chitosan-zinc oxide complex during chitin deacetylation. Carbohydr Polym 83:1565–1569
13. Limam Z, Selmi S, Sadok S et al (2011) Extraction and characterization of chitin and chitosan from crustacean by-products: biological and physicochemical properties. Afr J Biotechnol 10:640–647
14. Griffiths PR (2006) Introduction to the theory and instrumentation for vibrational spectroscopy. Wiley, Hoboken
15. Caldwell JD, Glembocki O, Bezares FJ et al (2011) Plasmonic nanopillar arrays for large-area, high-enhancement surface-enhanced Raman scattering sensors. ACS Nano 5:4046–4055
16. Chen Z, Zhan P, Wang ZL et al (2004) Two- and three-dimensional ordered structures of hollow silver spheres prepared by colloidal crystal templating. Adv Mater 16:417–422
17. Omura Y, Renbutsu E, Morimoto M et al (2003) Synthesis of new chitosan derivatives and combination with biodegradable polymer. Polym Adv Technol 14:35–39
18. Zabetakis D, Dressick WJ (2009) Selective electroless metallization of patterned polymeric films for lithography applications. ACS Appl Mater Interfaces 1:4–25
19. Renbutsu E, Okabe S, Omura Y et al (2007) Synthesis of UV-curable chitosan derivatives and palladium (II) adsorption behavior on their UV-exposed films. Carbohydr Polym 69:697–706
20. Smoukov SK, Bishop KJM, Campbell CJ et al (2005) Freestanding three-dimensional copper foils prepared by electroless deposition on micropatterned gels. Adv Mater 17:751–755
21. Renbutsu E, Okabe S, Omura Y et al (2008) Palladium adsorbing properties of UV-curable chitosan derivatives and surface analysis of chitosan-containing paint. Inter J Biol Macromol 43:62–68
22. Bao ZH, Weatherspoon MR, Shian S et al (2007) Chemical reduction of three-dimensional silica micro-assemblies into microporous silicon replicas. Nature 446:172–175
23. Chen Y, Zang XN, Gu JJ et al (2011) ZnO single butterfly wing scales: synthesis and spatial optical anisotropy. J Mater Chem 21:6140–6143
24. Liu XY, Zhu SM, Zhang D et al (2010) Replication of butterfly wing in TiO_2 with ordered mesopores assembled inside for light harvesting. Mater Lett 64:2745–2747
25. Song F, Su HL, Chen JJ et al (2011) Bioinspired ultraviolet reflective photonic structures derived from butterfly wings (*Euploea*). Appl Phy Lett 99:163705
26. Zhu S, Liu X, Chen Z et al (2010) Synthesis of Cu-doped WO_3 materials with photonic structures for high performance sensors. J Mater Chem 20:9126–9132
27. Peng WH, Zhu SM, Wang WL et al (2012) 3D network magnetophotonic crystals fabricated on *Morpho* butterfly wing templates. Adv Funct Mater 22:2072–2080
28. Wang J, Chen X, Wang G et al (2002) Melting behavior in ultrathin metallic nanowires. Phys Rev B 66:085408
29. Lisiecki I, Sack-Kongehl H, Weiss K et al (2000) Annealing process of anisotropic copper nanocrystals. Langmuir 16:8807–8808
30. Dippel M, Maier A, Gimple V et al (2001) Size-dependent melting of self-assembled indium nanostructures. Phys Rev Lett 87:095505
31. Link S, Wang ZL, El-Sayed MA (2000) How does a gold nanorod melt? J Phys Chem B 104:7867–7870
32. Vincendon M (1997) Regenerated chitin from phosphoric acid solutions. Carbohydr Polym 32:233–237
33. Lakhtakia A, Martin-Palma RJ, Motyka MA et al (2009) Fabrication of free-standing replicas of fragile, laminar, chitinous biotemplates. Bioinspir Biomim 4:034001
34. Weatherspoon MR, Cai Y, Crne M et al (2008) 3D rutile titania-based structures with *Morpho* butterfly wing scale morphologies. Angew Chem Int Ed Engl 47:7921–7923

Chapter 4
Surface-Enhanced Raman Scattering (SERS) Performance of Metal Scale Replicas

4.1 Introduction

Surface-enhanced Raman scattering (SERS) is one of the most important frontiers of nanostructured metals. Like IR spectroscopy, Raman spectroscopy is a basic and powerful tool to study molecular configurations. This technique can obtain similar but complementary information to IR spectroscopy, especially with the rapid development of laser technology. However, Raman spectrum is not easy to acquire as the Raman scattered beam is usually weak. Thus, SERS has attracted widespread attention ever since Fleischman et al. discovered SERS phenomenon in 1974 [1–6]. In a SERS process, Raman signals of molecules absorbed on supporting metal surfaces (especially on Ag, Au, and Cu) can be significantly amplified. The enhancement factor can be as high as 10^{10} [7]. To date, SERS has been widely applied in biomedical detection, environmental engineering, social security, chemistry and chemical engineering, and many other fields. For example, in biological sciences, SERS techniques can be used to study immune response and detect DNA/RNA for biological systems [8–10]. In medical research, SERS can help to study the interactions between anticancer drugs and DNA/proteins, in vivo detect tumors, and monitor virus genes or biological molecules associated with diseases, etc. [11, 12]. In environmental fields, SERS can be applied to monitor wastewater, waste gas, pesticide residues, toxic gas, and some other pollutants, etc. [13–15]. Moreover, SERS is also suitable to the detection of food additives, explosives, and some other threats to social safeties [16–18].

However, restricted by the preparation route, there is not an efficient way to mass-produce SERS substrates with high reproducibility. For example, Klarite™ of Renishaw is one of the most famous commercial SERS substrates at present. Its price is about $ 35 per single chip. Since SERS substrates are consumable, common laboratories and customers cannot afford such a high cost. This situation seriously prevents bringing the state-of-art results of SERS to human society. It is thus urgent to find some simple preparation methods to develop high-performance SERS

J. Gu et al., *Metallic Butterfly Wing Scales,* SpringerBriefs in Materials,
DOI 10.1007/978-3-319-12535-0_4

substrates with high sensitivity in detection, high reproducibility in performance, and low costs.

At present, the preparation of satisfactory SERS supports is a hot topic. Researchers have developed many high-performance SERS supports by optimizing metal microstructures or nanoparticle (NP) size and shapes [19–21]. It is generally believed that SERS performance can be adjusted by the microstructures of metal surface. In general, ways to prepare these structures can be divided into two groups. One contains bottom-up approaches, which are chemical routes to self-assemble building blocks like nanoparticles, nanowires, and nanorods into desired microstructures [22–24]. However, structures generated via chemical self-assembly are limited in topology and synthesizing large-area SERS structures in a bottom-up scheme is also difficult [25]. These shortcomings prevent the as-prepared SERS substrates from optimized structures and more powerful performance. The other includes a broad range of top–down approaches, which can be used to build more complicated structures. However, top–down fabrications need expensive instruments, complex processes, and long-processing time [26–29]. Thus, it is difficult to deploy these approaches in ordinary laboratories. Additionally, top–down methods usually use lights or particle flows to machine sample surfaces. Although these driving forces can easily generate 2D textures, it is difficult to obtain complex 3D structures because of the well-known "shading effects." Since substrates with 3D surface morphologies can bring more effective "hotspots" for SERS applications, simply using top–down approaches might be unsuitable to obtain high-performance and low-cost SERS substrates.

In this chapter, the SERS performance of Au butterfly scales is evaluated by detecting Raman signals of rhodamine 6G (R6G) molecules absorbed on Au scales. Influence exerted by various biostructures on SERS performance are discussed in detail.

4.2   Sample Preparation for SERS Detections

R6G molecules and crystal violet (CV) molecules (Fig. 4.1) have large Raman scattering cross-sections and have strong bonding tendencies with metal NPs. Therefore, R6G and CV molecules are usually chosen as ideal probes to evaluate SERS performance of metal supports [30, 31]. Herein, we use these two chemicals as analytes to study the SERS performance of metal butterfly scales. The SERS

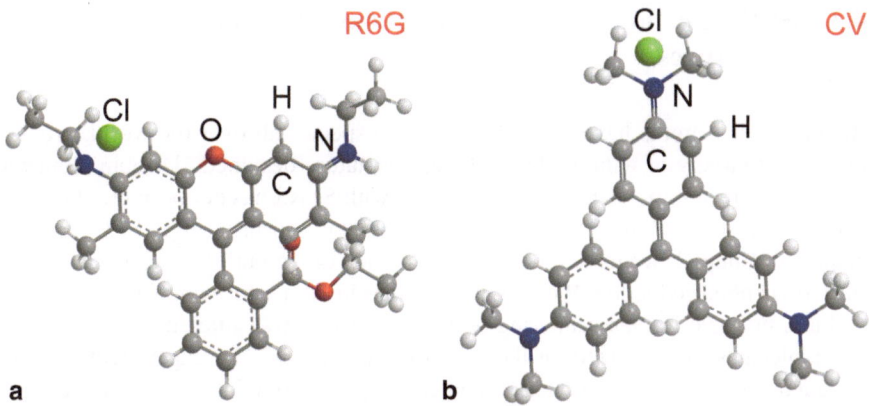

Fig. 4.1 Structural formula of Rhodamine 6G (R6G) (*left*) and crystal violet (CV) (*right*)

sample preparation and the related measurement parameters are described as follows:

1. R6G ethanol solutions and CV ethanol solutions with different concentrations were first prepared and kept in sealed brown bottles.
2. A metal butterfly wing as substrate was placed on a silicon wafer in a petri dish. R6G or CV solution with specific concentration was dripped onto wing surface and dried naturally in air.
3. The metal wing substrate bearing analytes was put onto the sample stage of a Raman spectrometer (Fig. 4.2). The sample was first found under a × 10 objective lens, and then measured under a × 50 short focal lens (numerical aperture: 0.75). The laser for excitation was 514.5 nm in wavelength and 2.0 mW in power. The integration time was 10 s and each spectrum was accumulated twice. Considering the repeatability of substrates in performance, Raman spectra on each sample were collected from different locations for several times to ensure the reliability of the measurement.

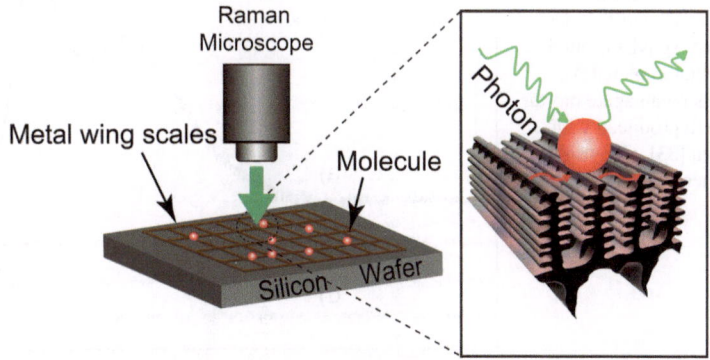

Fig. 4.2 Experimental setup for Raman measurements on 3D metal wing scales. (Reproduced with permission [32]. Copyright 2012, Wiley-VCH)

4.3   SERS Performance of Ag Replicas with Various Structures

Butterflies and moths have more than 174,500 species all over the world, providing a great variety of wing scales with various nature-designed 3D submicrometer structures. Since how, scale structures couple with SERS has not been clarified yet, it is difficult to determine which structure in metal will bring high SERS performance. In Chap. 2, we have introduced two kinds of Ag butterfly wing scales fabricated via photoreduction. We will discuss their SERS performance in this section.

First, intrinsic Raman scattering on bare substrates was studied. There were no dye molecules immobilized on these samples (Fig. 4.3). The experimental group included two types of Ag replicas prepared via photoreduction. In comparison, the control group involved a substrate formed by mechanically grinding an Ag wing replica in liquid N_2 and an original butterfly wing. No Raman signals were detected on these samples except on the organic original wing. This result indicates that the remaining chitin in Ag butterfly wings has little influence on SERS detection. Thus, these metal replicas can be considered "clean" to SERS detection.

Figure 4.4 compares SERS spectra of a R6G solution (10^{-3} M) dripped and dried on four different Ag substrates. These samples included a commercial Ag film (Substrate I, with an average particle size of~ 100 nm, from Alfa Aesar), a substrate of as-ground Ag wing replica (Substrate II), a Ag scale replica with the honeycomb structure of *P. paris* (Substrate III), and a Ag scale replica with the periodic structure replicated from *E. mulciber* (Substrate IV). The Raman peaks at 1649, 1597, 1571, 1533, 1508, 1364, and 1508 cm^{-1} can be attributed to C=C stretching vibration modes of the benzene ring, while those peaks below 1200 cm^{-1} indicate deformation vibration modes inside and outside of the plane of benzene ring [34–36]. Detailed characteristic Raman shifts of R6G are summarized in Table 4.1.

Fig. 4.3 Raman spectra of an original butterfly wing (*black*), Ag replicas of *E. mulciber* (*red*) and *P. pairs* (*blue*), and Ag NPs ground from the biotemplated Ag replicas using an agate mortar (*green*). (Reproduced with permission [33]. Copyright 2011, American Chemical Society)

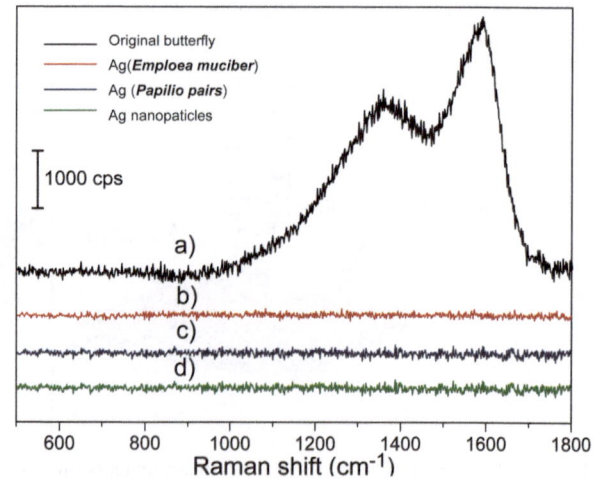

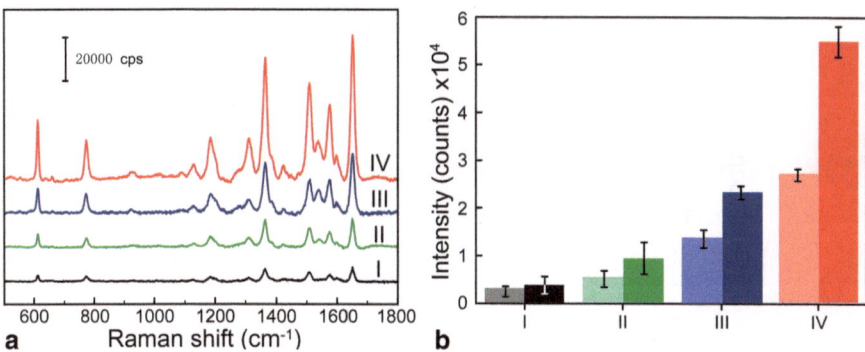

Fig. 4.4 a Surface-enhanced Raman scattering (SERS) spectra for a R6G solution (10^{-3} M) dripped on Substrate I (a smooth Ag *thin* film), Substrate II (Ag NPs ground from Ag wing replicas), Substrate III (Ag replica with the quasi-periodic submicrometer structures replicated from *P. paris*, and Substrate IV (Ag replica with the periodic submicrometer structures replicated from *E. mulciber*). **b** Comparison of the SERS signals at 612 (*left column*) and 1364 cm^{-1} (*right column*) from R6G for the four substrates in **a**. (Reproduced with permission [33]. Copyright 2011, American Chemical Society)

Table 4.1 Raman shifts of R6G

Raman shift (cm^{-1})	Assignment
1649	Arom C–C str
1597	Arom C–C str
1574	Arom C–C str
1533	Arom C–C str
1509	Arom C–C str
1364	Arom C–C str
1309	Arom C–C str
1190	Arom C–C str
1128	C–H in plane bend
773	C–H out-of-plane bend
634	C–C–C ring in plane bend
612	C–C–C ring in plane bend

Drastic improvements in the Raman scattering were achieved on Substrate IV with 3D periodic submicrometer structures, as compared with three other substrates (Fig. 4.4a). Figure 4.4b compares the peak intensities at 612 and 1364 cm^{-1}, respectively, which were extracted from the spectra acquired on the four substrates. Under the same experimental conditions, the intensities of R6G (10^{-3} M) Raman signals on Substrate IV were approximately 15, 6, and 2 times of magnitude higher than those on Substrates I–III, respectively. This result indicates that the structure and morphology of SERS substrates have great influence on the enhancement of Raman signals. Thus, there must be specific coupling relationship between scale structures and incident lights.

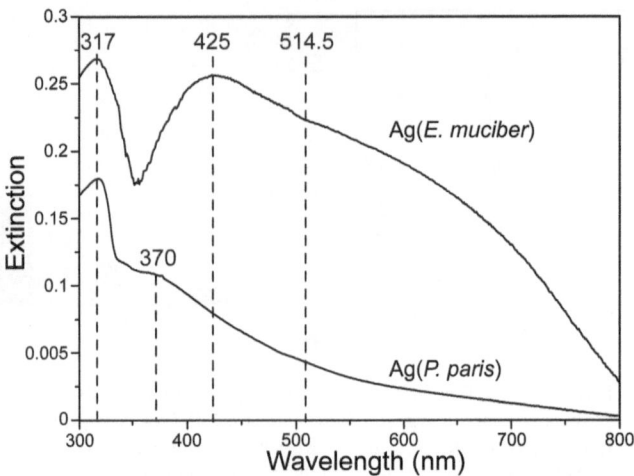

Fig. 4.5 Extinction spectra of Ag replicas of two butterfly species. The *upper* spectrum is from an *E. muciber* replica, while the *lower* one is from a *P. paris* replica. (Reproduced with permission [33]. Copyright 2011, American Chemical Society)

The extinction spectra of Substrate III and IV were further measured. As shown in Fig. 4.5, the peak at 317 nm appears for both Ag replicas. In comparison, a week peak at 370 nm appeares only for the Ag replica of *P. paris* (Substrate III), while a much broader and stronger peak at 425 nm exists only for the Ag replica of *E. muciber* (Substrate IV). Since the average Ag grain size was comparable for both samples, such difference in extinction spectra should originate from the various scale textures. It is generally believed that when an excitation wavelength approches to the extinction peak of a supporting material, the incident light will couple with the surface plasmon polaritons of substrates [31, 37]. This process will enhance the Raman intensities of analytes. Here, the extinction intensity at 514.5 nm on silver *E. muciber* was much stronger than that on silver *P. Paris*, suggesting that a stronger coupling was aroused by the 514.5 nm excitation on silver *E. muciber*. Thus, Substrate IV with periodic 3D structures is more effective in amplifying Raman signals of the molecules to be probed.

Moreover, recent studies indicated that metallic nanogratings or grooves have polarization effects on Raman scattering processes [38–40]. We thus studied the orientation relationship between Ag replicas and R6G Raman signals. Figure 4.6 shows the Raman spectra of R6G immobilized on two Ag replicas of *E. mulciber* with different orientations. Although some previous studies observed polarization phenomena originating from anisotropic metal nanostructures [38–40], Raman signals collected from the marked rectangular area (consisting of main ridges and struts) exhibited few differences in response to orientation. Two reasons might be responsible for this result. One is that the spacing between adjacent ridges and adjacent struts are ~720 and ~380 nm, respectively, much larger than the dimensions reported in previous works. The other is that the rectangle formed by main ridges and struts are actually not a planar pattern. The main ridges are about several

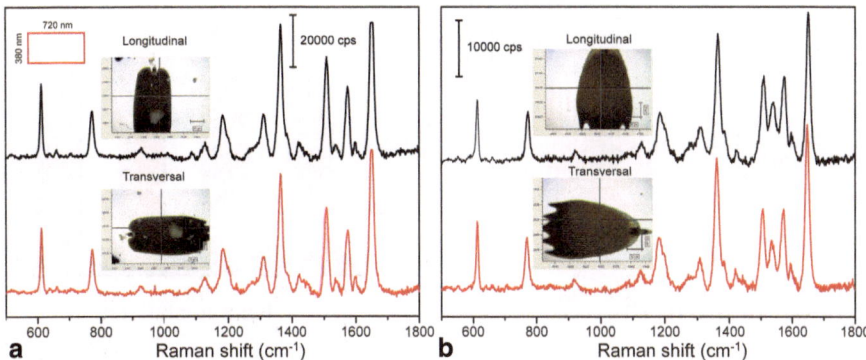

Fig. 4.6 Surface-enhanced Raman scattering (SERS) spectra of R6G acquired on the same Ag scale replica with different orientations. **a** *E. mulciber* and **b** *P. pairs*. The *rectangle* in **a** schematically illustrates the basic "window" structure of the *E. mulciber* scales shown in Fig. 2.4c. (Reproduced with permission [33]. Copyright 2011, American Chemical Society)

hundred nanometers higher than the struts, while the struts are ~200 nm in depth. Figure 4.6b provides the SERS spectra of an Ag replica of *P. pairs* in two orthogonal orientations. Similarly, there was no significant difference in Raman enhancement for different orientations. These experimental results indicate that metal scales show weak polarization effects on Raman scattering. Such performance is preferable as there will be no need to consider the substrate orientation in real applications.

Moreover, sensitivity in Raman detection and reproducibility in Raman signals are always two important figures of merits for a SERS substrate. Solutions of R6G with different concentrations were used to evaluate the detection sensitivity of Substrates III and IV. Figure 4.7 demonstrates the Raman intensity at 1364 cm^{-1} against the R6G concentration. The Raman intensity was stronger on Substrate IV than on Substrate III. Even a R6G trace as low as 10^{-9} M in concentration (with an estimated molecule coverage of ~10^{11}/cm^2 on the substrate) was detectable on Substrate IV. This detectable concentration achieved on Substrate IV was three orders

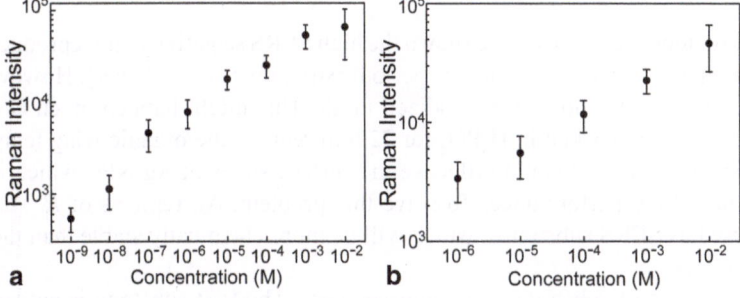

Fig. 4.7 Raman signals at 1364 cm^{-1} against different R6G concentrations. These data were collected on Ag replicas of **a** *E. mulciber* and **b** *P. paris*. (Reproduced with permission [33]. Copyright 2011, American Chemical Society)

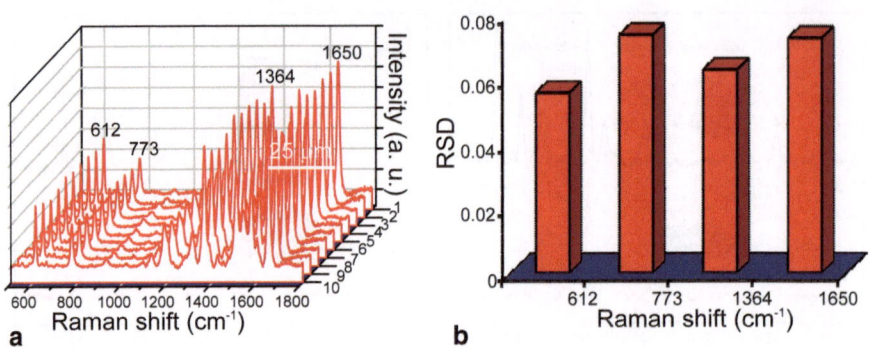

Fig. 4.8 a Raman spectra of R6G molecules on Ag replica with 3D periodic structures (Substrate IV). **b** Corresponding relative standard deviation (*RSD*) values at peak 612, 773, 1364, and 1650 cm^{-1}, respectively. (Reproduced with permission [33]. Copyright 2011, American Chemical Society)

of magnitude lower than that on Substrate III. Such a phenomenon should mainly originate from structural effects as these two scales have comparable effective areas to assemble Ag crystallites. Figure 4.8 shows the consistency of ten SERS spectra acquired from various spots on Substrate IV. The relative standard deviation relative standard deviation (RSD) in peak intensity was lower than 8%, indicating an excellent reproducibility of Raman signals on Substrate IV (Fig. 4.8b).

In summary, Ag scale replicas with a 3D periodic structure has much higher sensitivity than structureless metal NPs in detecting R6G molecules. The coupling effect between light and long-range-ordered scale structures may effectively amplify the Raman signals of analytes. In addition, analytes' Raman spectra acquired on these periodic metal structures exhibit high reproducibility as well.

4.4   SERS Properties of Au Butterfly Wing Scales with Ordered Structures

In the previous section, we have shown the high SERS sensitivity and reproducibility of Ag butterfly wing scales with 3D periodic structures (*E. mulciber*). However, Ag NPs are not stable and can be oxidized in air. This might happen much easier as Ag replicas were soaked in H$_3$PO$_4$ for 72 h to remove the organic wing templates. The oxidation process could influence the surface states of Ag NPs, which in turn affect the SERS performance. To solve this problem, Au replicas of *E. mulciber* were used as SERS substrates, which will be more chemically stable than their Ag counterparts.

Three types of substrates are compared here. The first substrate is an Au scale replica prepared with a simple PVD process and will be referred as "physical scale" hereafter. The second one is a commercial SERS substrate (Klarite™) . This chip is formed by coating Au NPs on a specific silicon photonic structure, which comprises numerous regularly arranged small holes. Based on such an architecture, Klarite™

chips can effectively control their surface plasmons to amplify Raman signals. The last substrate is an Au replica prepared via the electroless deposition approach introduced in Chap. 3 (referred as "chemical scale" hereafter).

R6G solutions of different concentrations were dripped and dried on these three Au substrates. Figure 4.9 shows the corresponding Raman signals of R6G. It should be noted that once using Au butterfly wing scales as SERS substrates, the detectable

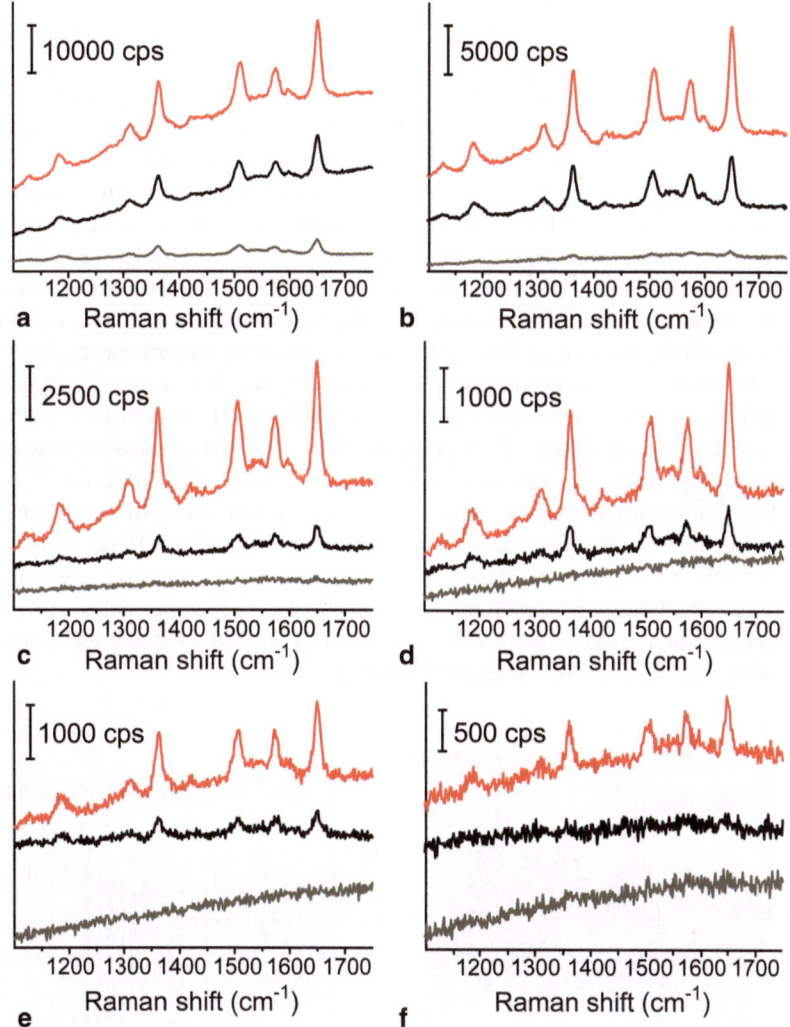

Fig. 4.9 a–f Comparison of Raman signals from R6G on three Au surface-enhanced Raman scattering (SERS) substrates. R6G concentrations: **a** 10^{-6}, **b** 10^{-8}, **c** 10^{-10}, **d** 10^{-11}, **e** 10^{-12}, and **f** 10^{-13} M. Data from *top to bottom* in each panel were collected on Au butterfly scales chemically synthesized, a commercial SERS substrate (Klarite™), and Au scales prepared through a physical deposition process. (Reproduced with permission [42]. Copyright 2011, Wiley-VCH)

lower limit of analyte concentration can surprisingly decrease at least by one order of magnitude (R6G, 10^{-13} M), as compared with that on Klarite^{TM} (10^{-12} M) and on other SERS substrates reported in recent literatures [30, 31, 41]. Moreover, the lowest detectable R6G concentration on chemical scales was four orders of magnitude lower than that (10^{-9} M) on physical scales. Although there is currently no direct experimental evidence proving that the chemical scales can be directly used in single-molecule detection, the number of molecules within the detected area was roughly estimated using a reported method [30]. One millileter of R6G ethanol solution (10^{-13} M) contains $\sim 6 \times 10^7$ R6G molecules. It was dripped onto the chemical scale surface and the liquid drop could finally spread out into an area of ~ 1 cm^2. The cross-section of the laser beam applied for Raman detection was ~ 2 μm in diameter. Suppose these 6×10^7 R6G molecules were adsorbed evenly on the ~ 1 cm^2 area, the collected Raman signals should come from less than ten molecules [42].

As mentioned above, reproducibility in performance is another important criterion to evaluate a SERS substrate. Recently, lots studies focused on the reproductivity of SERS substrates in addition to high sensitivity [27, 30, 41]. Accidental factors are inevitable to the bottom-up methods, giving rise to performance varieties for samples from different batches. Though substrates with high reproductivity and stability have been fabricated based on some top–down approaches such as plasma etching and nano-imprint lithography [43, 44], these methods depend on high-tech equipment and are comparatively low efficient. In comparison, butterfly wings are designed by nature. Their structures have been determined by gene-engineering and optimized by natural selection for over hundreds of millions of years. This fact grants butterfly scales high repeatability in functions. In Fig. 4.10a, the Raman signals were measured at 30 randomly chosen spots from 10 Au scales. The RSD is 5.2%, comparable to [30, 45] or even better than [41, 46, 47] those reported in very recent studies. This result indicates that Au butterfly wing scales with gene-engineered structures have excellent reproducibility in Raman signal enhancement and can be applied in laboratories and industries.

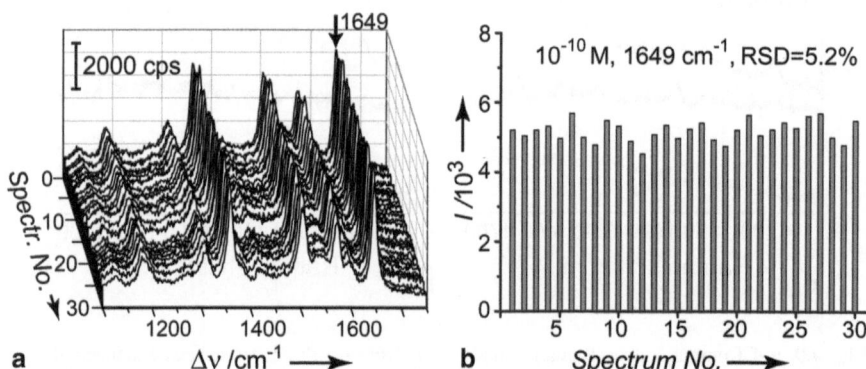

Fig. 4.10 **a** Raman signals (R6G, 10^{-10} M) obtained at 30 different spots from 10 Au chemical scales. **b** Raman intensities at 1649 cm^{-1} in **a**. (Reproduced with permission [42]. Copyright 2011, Wiley-VCH)

To study whether these substrates can be mass-produced or not, SERS substrates based on whole butterfly wings were prepared. Raman spectra acquired on a whole Au wing (*E. mulciber*) as SERS substrate are provided in Fig. 4.11. Obviously, dilute R6G solution (10^{-13} M) is detectable on this specimen. This result indicates that large-area SERS substrates based on a whole wing have high sensitivity in detecting chemicals as well. Most importantly, SERS substrates for chemical detection are consumables, while present commercial substrates are quite expensive as mentioned before. A consumable SERS substrate of 4×4 mm^2 in size costs ~$ 35(Klarite^{TM}) , which most labs and consumers are not able to afford.

This situation drastically limits broader applications of SERS technologies. However, the cost (~$ 2.5) of a Au wing tailored into 4×4 mm^2 is about one order of magnitude cheaper than a Klarite^{TM} chip. Such a wing with more than 100,000 single scales can be prepared within 13 h without the removal of original chitin in a

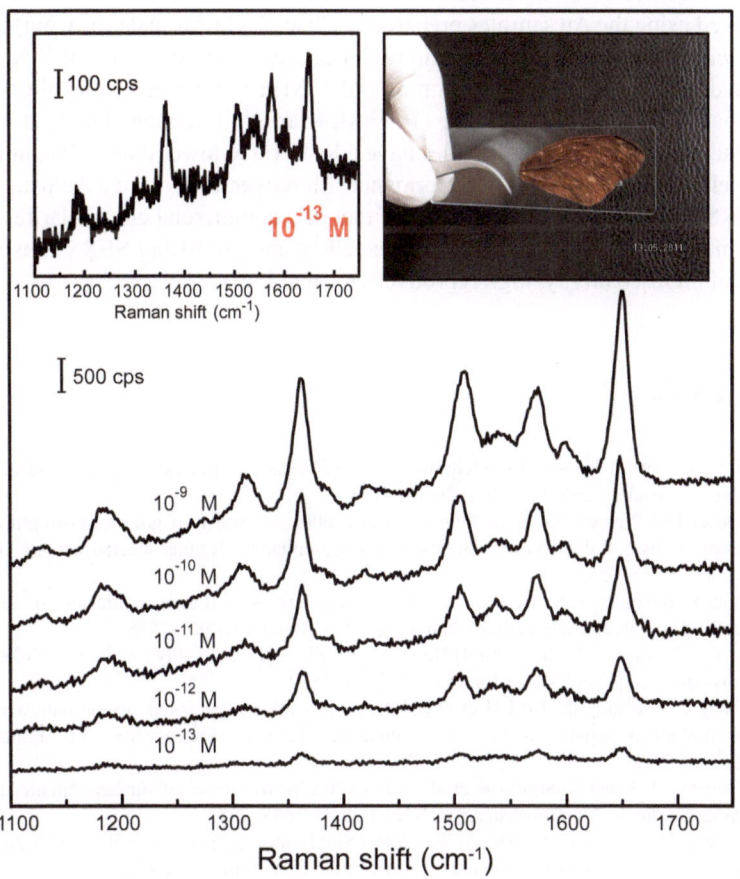

Fig. 4.11 Surface-enhanced Raman scattering (SERS) performance of a whole Au butterfly wing (without the removal of original biotemplate). Raman signals are from R6G molecules immobilized on the Au wing. (Reproduced with permission [42]. Copyright 2011, Wiley-VCH)

common laboratory. These substrates are thus promising to make ultrasensitive Raman analysis a *daily* diagnostic method for chemists, life scientists, and physicists, with a detection limit close to single-molecule-level and an excellent reproducibility in performance.

4.5 Summary

As demonstrated above, Raman signals from R6G on various metal scales exhibit great difference. This phenomenon originates from the different structural characteristics of natural butterfly scales. A R6G solution with a concentration down to 10^{-9} M can be identified on Ag replicas of *E. mulciber* with 3D periodic structures. This detection limit is three orders of magnitude lower than that on Ag replicas with quasi-ordered textures inherited from *P. paris*. Such performance can be further improved using the Au samples prepared in Chap. 3. On Au scale (or wing) replicas of *E. mulciber*, a dilute R6G solution with concentration down to 10^{-13} M can be detected. This detectable concentration (10^{-13} M) is ten times of magnitude lower than that on commercial substrate (10^{-12} M, Klarite^{TM}). Raman signals at 30 spots randomly chosen from 10 Au scales have a RSD value lower than 5.2 %, indicating an excellent reproducibility in performance. Moreover, the cost of a Au wing replica as SERS substrate is about one tenth of that of a commercial chip (Klarite^{TM}) . All these merits make Au butterfly scales excellent and promising SERS substrates in terms of high sensitivity, high reproductivity, and low cost.

References

1. Fleischmann M, Hendra PJ, Mcquillan AJ (1974) Raman spectra of pyridine adsorbed at a silver electrode. Chem Phys Lett 26:163–166
2. Tessier PM, Velev OD, Kalambur AT et al (2000) Assembly of gold nanostructured films templated by colloidal crystals and use in surface-enhanced Raman spectroscopy. J Am Chem Soc 122:9554–9555
3. Chan S, Kwon S, Koo TW et al (2003) Surface-enhanced Raman scattering of small molecules from silver-coated silicon nanopores. Adv Mater 15:1595–1598
4. Cerf A, Molnar G, Vieu C (2009) Novel approach for the assembly of highly efficient SERS substrates. ACS Appl Mater Interfaces 1:2544–2550
5. Zhang BH, Wang HS, Lu LH et al (2008) Large-area silver-coated silicon nanowire arrays for molecular sensing using surface-enhanced Raman spectroscopy. Adv Funct Mater 18:2348–2355
6. Katayama I, Koga S, Shudo K et al (2011) Ultrafast dynamics of surface-enhanced Raman scattering due to Au nanostructures. Nano Lett 11:2648–2654
7. Blackie EJ, Le Ru EC, Etchegoin PG (2009) Single-molecule surface-enhanced Raman spectroscopy of nonresonant molecules. J Am Chem Soc 131:14466–14472
8. Kang T, Yoo SM, Yoon I et al (2010) Patterned multiplex pathogen DNA detection by Au particle-on-wire SERS sensor. Nano Lett 10:1189–1193

9. Hu JA, Zhang CY (2010) Sensitive detection of nucleic acids with rolling circle amplification and surface-enhanced Raman scattering spectroscopy. Anal Chem 82:8991–8997

10. Bell SEJ, Sirimuthu NMS (2006) Surface-enhanced Raman spectroscopy (SERS) for sub-micromolar detection of DNA/RNA mononucleotides. J Am Chem Soc 128:15580–15581

11. Papadopoulou E, Bell SEJ (2011) DNA reorientation on Au nanoparticles: label-free detection of hybridization by surface enhanced Raman spectroscopy. Chem Commun 47:10966–10968

12. Prado E, Daugey N, Plumet S et al (2011) Quantitative label-free rna detection using surface-enhanced Raman spectroscopy. Chem Commun 47:7425–7427

13. Wang XT, Shi WS, She GW et al (2010) High-performance surface-enhanced Raman scattering sensors based on Ag nanoparticles-coated Si nanowire arrays for quantitative detection of pesticides. Appl Phy Lett 96:053104

14. Ruan CM, Luo WS, Wang W et al (2007) Surface-enhanced Raman spectroscopy for uranium detection and analysis in environmental samples. Anal Chim Acta 605:80–86

15. Li XH, Chen GY, Yang LB et al (2010) Multifunctional Au-coated TiO_2 nanotube arrays as recyclable SERS substrates for multifold organic pollutants detection. Adv Funct Mater 20:2815–2824

16. Yang LB, Ma LA, Chen GY et al (2010) Ultrasensitive SERS detection of TNT by imprinting molecular recognition using a new type of stable substrate. Chem-Eur J 16:12683–12693

17. Lee S, Choi J, Chen L et al (2007) Fast and sensitive trace analysis of malachite green using a surface-enhanced Raman microfluidic sensor. Anal Chim Acta 590:139–144

18. Lopez-Tocon I, Otero JC, Arenas JF et al (2010) Trace detection of triphenylene by surface enhanced Raman spectroscopy using functionalized silver nanoparticles with *bis*-acridinium lucigenine. Langmuir 26:6977–6981

19. Kim MG, Lee SW, Lee KS et al (2011) Highly sensitive biosensing using arrays of plasmonic Au nanodisks realized by nanoimprint lithography. ACS Nano 5:897–904

20. Caldwell JD, Glembocki O, Bezares FJ et al (2011) Plasmonic nanopillar arrays for large-area, high-enhancement surface-enhanced Raman scattering sensors. ACS Nano 5:4046–4055

21. Tuan VD, Dhawan A, Du Y et al (2011) Hybrid top-down and bottom-up fabrication approach for wafer-scale plasmonic nanoplatforms. Small 7:727–731

22. Lim DK, Jeon KS, Hwang JH et al (2011) Highly uniform and reproducible surface-enhanced Raman scattering from DNA-tailorable nanoparticles with 1-nm interior gap. Nat Nanotechnol 6:452–460

23. Xia X, Zeng J, Mcdearmon B et al (2011) Silver nanocrystals with concave surfaces and their optical and surface-enhanced Raman scattering properties. Angew Chem Int Ed 50:12542–12546

24. Liberman V, Yilmaz C, Bloomstein TM et al (2010) A nanoparticle convective directed assembly process for the fabrication of periodic surface enhanced Raman spectroscopy substrates. Adv Mater 22:4298–4302

25. Ko H, Singamaneni S, Tsukruk VV (2008) Nanostructured surfaces and assemblies as SERS media. Small 4:1576–1599

26. Im H, Bantz KC, Lindquist NC et al (2010) Vertically oriented sub-10-nm plasmonic nanogap arrays. Nano Lett 10:2231–2236

27. Hatab NA, Hsueh CH, Gaddis AL et al (2010) Free-standing optical gold bowtie nanoantenna with variable gap size for enhanced Raman spectroscopy. Nano Lett 10:4952–4955

28. Qin Y, Pan AL, Liu LF et al (2011) Atomic layer deposition assisted template approach for electrochemical synthesis of Au crescent-shaped half-nanotubes. ACS Nano 5:788–794

29. Banholzer MJ, Millstone JE, Qin LD et al (2008) Rationally designed nanostructures for surface-enhanced Raman spectroscopy. Chem Soc Rev 37:885–897

30. Zhao Y, Zhang XJ, Ye J et al (2011) Metallo-dielectric photonic crystals for surface-enhanced Raman scattering. ACS Nano 5:3027–3033

31. Huang Z, Meng G, Huang Q et al (2010) Improved SERS performance from Au nanopillar arrays by abridging the pillar tip spacing by Ag sputtering. Adv Mater 22:4136–4139

32. Tan YW, Gu JJ, Xu LH et al (2012) High-density hotspots engineered by naturally piled-up subwavelength structures in three-dimensional copper butterfly wing scales for surface-enhanced Raman scattering detection. Adv Funct Mater 22:1578–1585

33. Tan YW, Zang XN, Gu JJ et al (2011) Morphological effects on surface-enhanced Raman scattering from silver butterfly wing scales synthesized via photoreduction. Langmuir 27:11742–11746

34. Wang HH, Liu CY, Wu SB et al (2006) Highly Raman-enhancing substrates based on silver nanoparticle arrays with tunable sub-10 nm gaps. Adv Mater 18:491–495

35. Baik JM, Lee SJ, Moskovits M (2009) Polarized surface-enhanced Raman spectroscopy from molecules adsorbed in nano-gaps produced by electromigration in silver nanowires. Nano Lett 9:672–676

36. Pristinski D, Tan S, Erol M et al (2006) In situ SERS study of rhodamine 6G adsorbed on individually immobilized Ag nanoparticles. J Raman Spectroscopy 37:762–770

37. Zuev VS, Frantsesson AV, Gao J et al (2005) Enhancement of Raman scattering for an atom or molecule near a metal nanocylinder: quantum theory of spontaneous emission and coupling to surface plasmon modes. J Chem Phys 122:214726

38. Nagasawa F, Takase M, Nabika H et al (2011) Polarization characteristics of surface-enhanced Raman scattering from a small number of molecules at the gap of a metal nano-dimer. Chem Commun 47:4514–4516

39. Jiao Y, Ryckman JD, Ciesielski PN et al (2011) Patterned nanoporous gold as an effective SERS template. Nanotechnology 22:295302

40. Lee S, Shin J, Lee Y-H et al (2010) Fabrication of the funnel-shaped three-dimensional plasmonic tip arrays by directional photofluidization lithography. ACS Nano 4:7175–7184

41. Choi D, Choi Y, Hong S et al (2010) Self-organized hexagonal-nanopore SERS array. Small 6:1741–1744

42. Tan Y, Gu J, Zang X et al (2011) Versatile fabrication of intact three-dimensional metallic butterfly wing scales with hierarchical sub-micrometer structures. Angew Chem Int Ed 50:8307–8311

43. Wu W, Hu M, Ou FS et al (2010) Cones fabricated by 3D nanoimprint lithography for highly sensitive surface enhanced Raman spectroscopy. Nanotechnology 21:255502

44. Weiss SM, Ryckman JD, Liscidini M et al (2011) Direct imprinting of porous substrates: a rapid and low-cost approach for patterning porous nanomaterials. Nano Lett 11:1857–1862

45. Kamińska A, Dzięcielewski I, Weyher JL et al (2011) Highly reproducible, stable and multiply regenerated surface-enhanced Raman scattering substrate for biomedical applications. J Mater Chem 21:8662–8669

46. Liu X, Linn NC, Sun C-H et al (2010) Templated fabrication of metal half-shells for surface-enhanced Raman scattering. Phys Chem Chem Phys 12:1379–1387

47. He D, Hu B, Yao Q-F et al (2009) Large-scale synthesis of flexible free-standing SERS substrates with high sensitivity: electrospun PVA nanofibers embedded with controlled alignment of silver nanoparticles. ACS Nano 3:3993–4002

Chapter 5
Surface-Enhanced Raman Scattering (SERS) Mechanisms of Metal Scale Replicas

5.1 Introduction

As shown in Chap. 4, surface-enhanced Raman spectroscopy (SERS) substrates with scale structure have unique advantages in terms of sensitivity, repeatability, and mass-producibility. However, the mechanism by which such performance is achieved should be clarified. This may help select appropriate biostructures from countless biological candidates for SERS application. In this chapter, we first discuss the SERS performance of differently textured metal scales, whose microstructures were regulated by changing the metal deposition time (DT). We then compare the contributions from different structural features and target the key contributor to SERS performance. Such a structure is then analyzed using a finite element method (FEM). The resulting mechanism will be checked by studying the SERS performance of Cu scales with different structures. All these results will illustrate a mechanism by which metal butterfly scale replicas can effectively enhance the Raman signals of analytes.

To date, the mechanism of SERS has not been fully clarified. Induced by interactions between radiations and rough metal surfaces [1], SERS is closely associated with the shape of metal particles and the surface-texturing of metal structures. However, molecular orientations, conformations, and bonding states on the surface can affect the enhancement process as well. Other active factors include the intensity, frequency, and polarization of the incident light. The tangle of these influences bring great complexities to SERS studies.

It is now generally believed that both physical and chemical factors can contribute to the SERS phenomenon [2, 3]. The Raman intensity I_{SERS} of a SERS process can be described by the following relation [4]:

© Jiajun Gu, Di Zhang, and Yongwen Tan 2015   69
J. Gu et al., *Metallic Butterfly Wing Scales,* SpringerBriefs in Materials,
DOI 10.1007/978-3-319-12535-0_5

$$I_{SERS} \propto \left[\left| \vec{E}(\omega_0) \right|^2 \left| \vec{E}(\omega_S) \right|^2 \right] \sum_{\rho,\sigma} \left| (\alpha_{\rho,\sigma})_{fi} \right|^2. \qquad (5.1)$$

where $E(\omega_0)$ and $E(\omega_S)$ are the intensities of local electric fields with frequencies of ω_0 for the incident light and ω_S for the scattered light, respectively. The electric field direction of the incident light and the Raman scattered light are denoted by ρ and σ respectively. $(\alpha_{\rho\sigma})fi$ is the polarizability tensor of an intermediate state between the initial and final state. The former part of I_{SERS} means that once the electric field intensity of either incident light or scattered light increases, the Raman intensity will be amplified as well. This physical contribution is widely believed to result from a coupling effect by the incident light and the excited surface plasmons. The latter part suggests that Raman intensity increases with $|(\alpha_{\rho\sigma})fi|$. This contribution originates from a chemical enhancement, which is sometimes considered as an increase in the polarizability of the molecules bound on metal surface. Also, there is another chemical enhancement theory named as charge transfer mechanism. It involves a charge transfer process between molecules and metal when the radiation wavelength is resonant with charge-transfer-states [5]. Although the chemical enhancement mechanisms seems more complicated than their physical counterpart, present experimental results suggest that the physical enhancement is dominant in SERS processes [1]. Thus, we mainly focus on the physical enhancement mechanism in this chapter.

The physical enhancement mechanism is also referred as electromagnetic (EM) mechanism. In this view, a strongest electromagnetic field provided by metal surface will occur once the surface plasmon frequency is resonant with the excitation wavelength [1]. The resulting localized surface plasmon resonance (LSPR) can significantly amplify the electric fields of both incident light and Raman scattered beams. It acts as an "antenna" that can amplify the intensity of the scattered light by an enhancement factor approximate proportional to $|E|^4$ [3]. Therefore, even a slight increase in the intensity of local electric field may cause a giant amplification in Raman signals. This is the reason why metal NPs with sharp edges or structures with nano "gaps" [6, 7] are beneficial to SERS, as they can effectively amplify local electric fields under radiations [8].

Based on the EM theory, SERS performance obviously depends on the supporting materials and their surface morphologies. In some cases, small changes in surface texturing may result in several orders of changes in Raman enhancement factor [9–12]. Recently, SERS substrates can be formed by chemical assembly of metal NPs with special shapes or core-shell structures. Also, they can be fabricated via photolithography [13–18]. All these methods tried to generate strong local electromagnetic fields, i.e., "hotspots," on metal surface. High-density "hotspots" can yield high SERS performance.

Compared to substrates with 2D textures, 3D nanostructured substrates are preferable as they can extend the hotspots to the additional dimension. Though many 3D structures have shown excellent enhancement factors in laboratories, they are still difficult to be commercialized [12, 19–21].

5.2   SERS Performance of Cu Butterfly Wing Scales

The biotemplates discussed here are the scales of *Euploea mulciber*. Figure 5.1a, b show optical images of a male *E. mulciber* butterfly and the scales on its forewing. As introduced in Chap. 2, the forewing of *E. mulciber* can reflect brilliant dark blue at a particular view angle to satisfy survival requirements including courtship, communication, and evading predators [22, 23]. Figure 5.1c, f show the scanning electron microscope (SEM) images of the blue butterfly wing scales with 3D periodic structures. Typical photonic structures such as main ridges, struts, and ribs, are denoted [25, 26].

Fig. 5.1 Morphology of original butterfly wing scales. **a** *E. mulciber*. **b** Optical microscopy image of wing scales in the rectangular shown in (**a**). **c** and **d** are the scanning electron microscope (SEM) images of natural blue scales in (**b**) with (**c**) *top view* and (**d**) *side view*. Insets are zoomed-up SEM images. Scales bars: 250 nm. **e** and **f** provide cross-section morphologies of scales. (Reproduced with permission [24] Copyright 2012, Wiley-VCH)

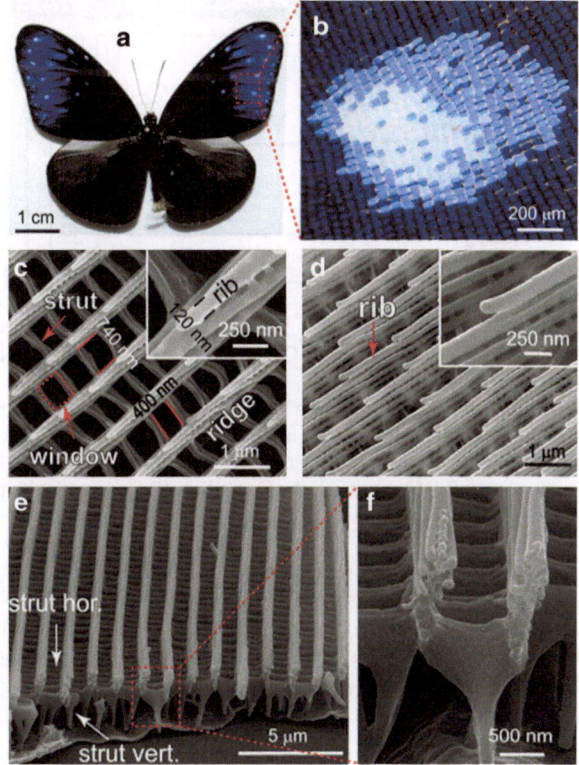

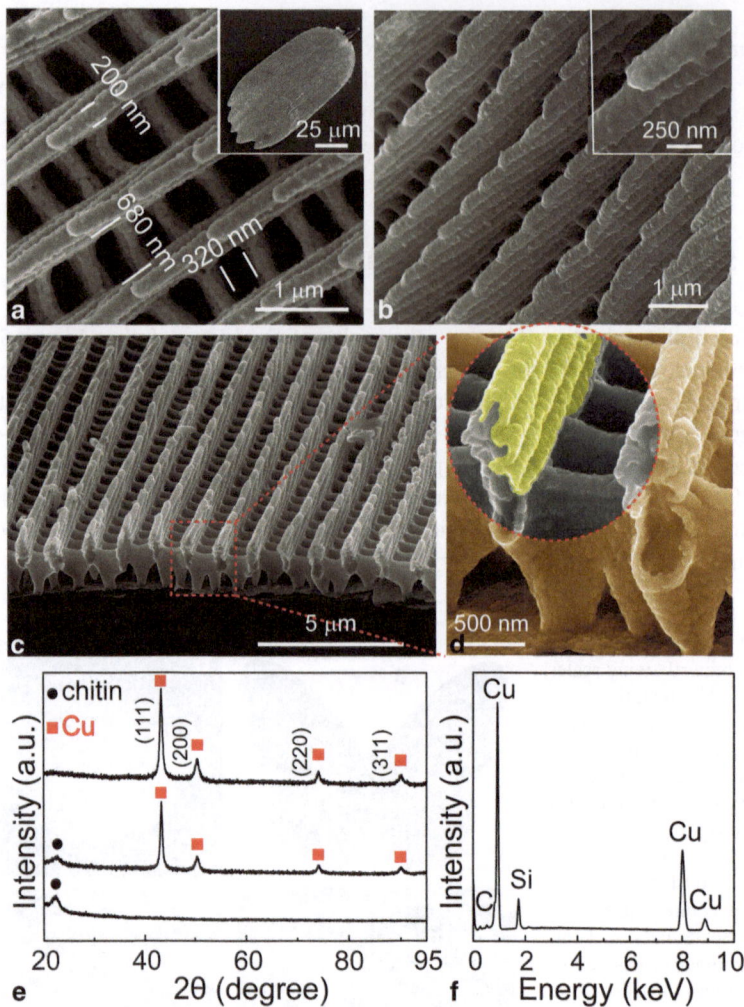

Fig. 5.2 a–d Scanning electron microscope (SEM) images of the Cu scale replica obtained via electroless Cu deposition for 10 min. **e** X-ray diffraction (XRD) results of the original wing, wings functionalized by Au NPs as catalysts, and Cu replicas, respectively. **f** Energy dispersive X-ray spectroscopy (EDS) results on the Cu replica. (Reproduced with permission [24] Copyright 2012, Wiley-VCH)

Cu wing replicas were prepared through the electroless deposition method introduced in Chap. 3. Figure 5.2 provides the SEM images of Cu scale replicas. The microstructures of the original butterfly wing scales (Fig. 5.1c) could be well-kept through the uniform deposition of Cu NPs (Fig. 5.2a). The Cu butterfly scales not only retained the main ridges of original scales at a subwavelength level but also well-kept the nanosized rib structures. Figure 5.2e gives the X-ray diffraction (XRD) results of the Cu butterfly wings. Obviously, after H_3PO_4 treatment,

Fig. 5.3 Raman spectra of **a** Cu NPs mechanically ground from Cu scale replicas, **b** Cu replicas with 3D periodic structures, and **c** original butterfly wing. (Reproduced with permission [24] Copyright 2012, Wiley-VCH)

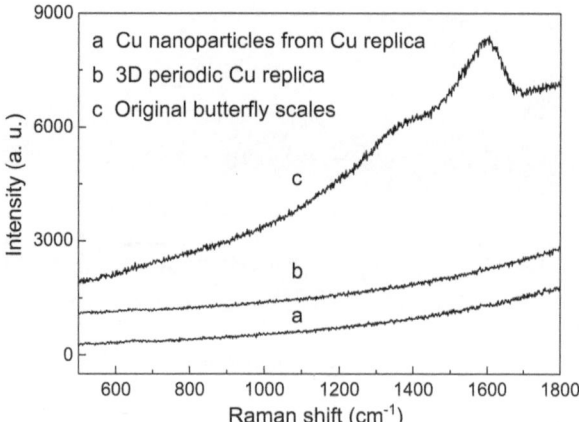

amorphous wing ingredients (chitin) were removed. The grain size of Cu is estimated to be ~ 16.9 nm, and the diffraction peaks are attributed to be (111), (200), (220), and (311) plane of *fcc* Cu, respectively. Figure 5.2f shows an energy dispersive X-ray spectroscopy (EDS) spectrum of the Cu wing, showing that the original bio-components were successfully converted to Cu.

Figure 5.3 compares the Raman spectra of Cu NPs ground from the Cu replicas, Cu replicas with 3D periodic structures, and an original butterfly wing, respectively. The Raman signals of chitin could be detected on the original butterfly wing, while no signals were found either on Cu replicas or on their as-ground NPs. This result indicates that most organic chemicals in Cu wings were melted off after the exposure in H_3PO_4. Most importantly, these Cu specimens do not produce intrinsic Raman signals and are suitable as SERS substrates.

Figure 5.4 provides the SERS spectra of R6G solution (10^{-5} M) dripped and dried on three different Cu substrates. Substrate I was composed of Cu NPs ground

Fig. 5.4 Surface-enhanced Raman Scattering (SERS) spectra of R6G (10^{-5} M) acquired on **a** Cu scale replicas deposited for 10 min, **b** Cu thin film formed on a silicon wafer via the same deposition process used in (**a**), and **c** Cu NPs mechanically ground from the Cu scale replicas in liquid N_2. (Reproduced with permission [24] Copyright 2012, Wiley-VCH)

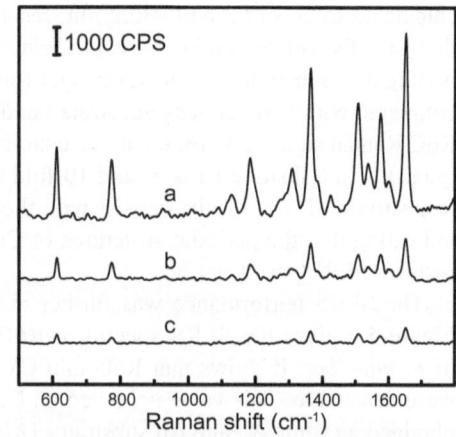

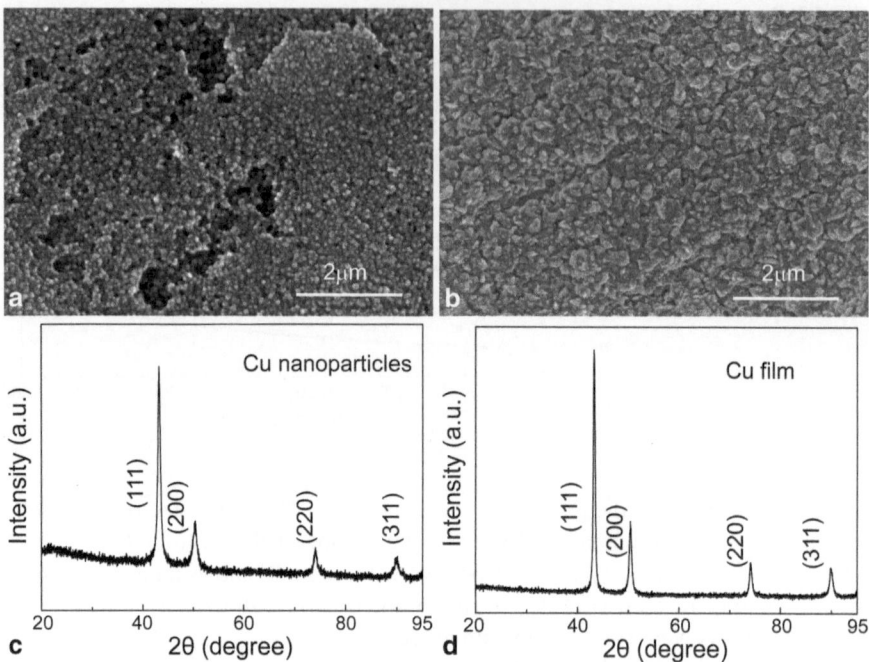

Fig. 5.5 Scanning electron microscope (SEM) images and X-ray diffraction (XRD) results of Cu powders used for SERS detection. **a** and **c** Cu NPs ground from a Cu replica in liquid N_2. **b** and **d** Cu thin film prepared on a silicon wafer via electroless Cu deposition. (Reproduced with permission [24] Copyright 2012, Wiley-VCH)

from Cu scales in liquid N_2, Substrate II was a Cu film formed by directly coating Cu NPs on a Si wafer, and Substrate III was a Cu wing with periodically structured scales. The same experimental conditions were applied to deposit Cu on all these substrates. Figure 5.5 shows SEM images and XRD results of Substrate I and II. Using the Scherrer equation, the average Cu grain size on Substrate I and II was calculated to be~17.2 and~30.2 nm, respectively. Since previous studies proved that Cu NPs with 50 nm in size could bring the best SERS effects [21], it is not surprising that Substrate II yielded stronger Raman signals than Substrate I. Moreover, compared with structureless Substrate I and II, Substrate III significantly enhanced R6G Raman signals. At 1650 cm^{-1}, a characteristic Raman peak of R6G, the Raman intensity on Substrate III is 5- and 10-fold stronger than that on Substrate I and II, respectively. These results consist with those obtained in Chap. 4 on Ag replicas, indicating that the periodic structures of Cu replica have obvious and positive effects on SERS performance.

The SERS performance was further measured on a whole Cu butterfly wing. Figure 5.6 gives the SERS spectra of R6G and CV on a whole Cu wing replica of *E. mulciber*. It shows that R6G and CV traces with a concentration of 10^{-8} M were detectable. This value achieved on Cu substrate is even comparable to those obtained on some Ag and Au substrates [27–29].

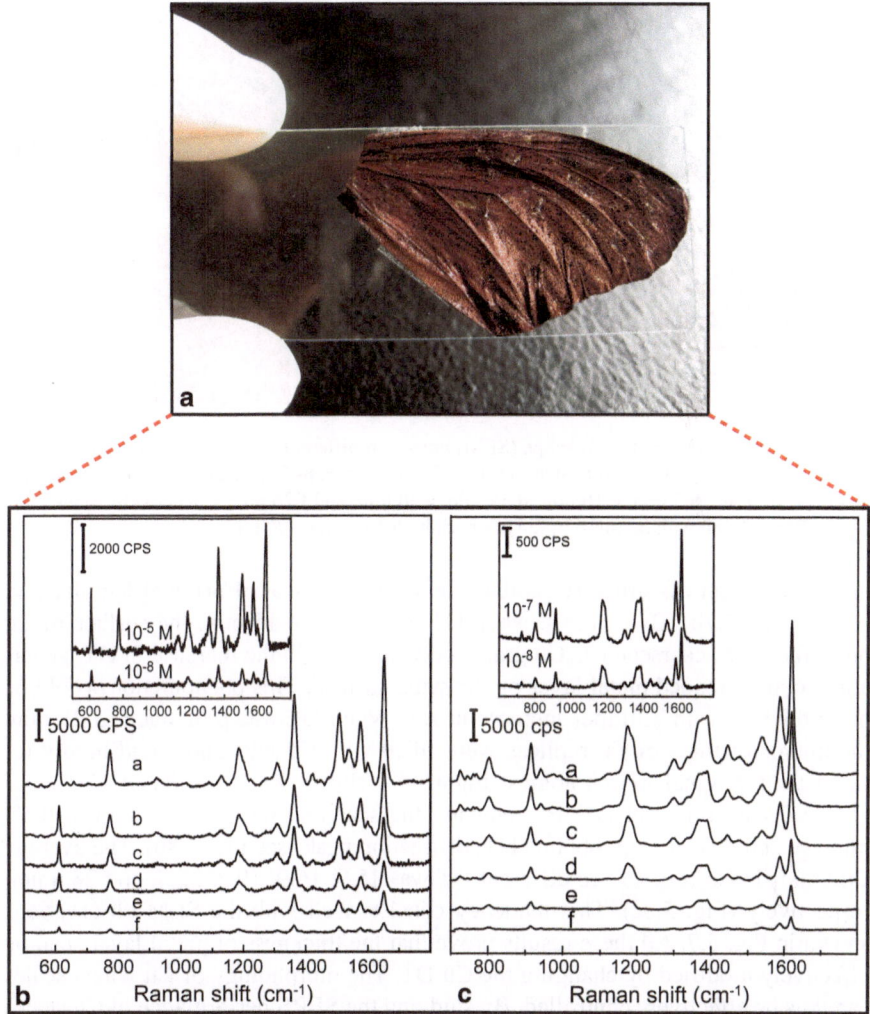

Fig. 5.6 Surface-enhanced Raman scattering (SERS) spectra of R6G and CV on Cu wing replicas of *E. mulciber*, with a concentration of **a** 10^{-3} M, **b** 10^{-4} M, **c** 10^{-5} M, **d** 10^{-6} M, **e** 10^{-7} M, and **f** 10^{-8} M, respectively. These replicas were prepared with a Cu deposition for ten minutes. (Reproduced with permission [24] Copyright 2012, Wiley-VCH)

5.3   Dominant Structural Contributor to SERS Performance

To clarify which microstructures of metal scales plays the key role in SERS, a series of Cu butterfly wing scales (see Fig. 5.7) were prepared by increasing the electroless DT from 5 to 25 min, with an interval of 5 min. Experimental results show that Cu butterfly scales prepared with 10 min in DT could best-keep the original biostructures (Fig. 5.7c). When DT was 5 min, the main ridges and rib structures of butterfly wings scales could not be uniformly covered by metal NPs. This made

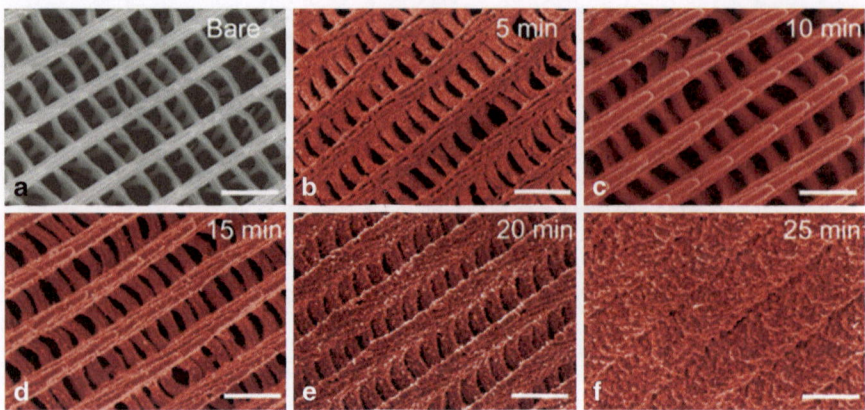

Fig. 5.7 Scanning electron microscope (SEM) images of different Cu scales prepared with different deposition time (DT). **a** Chitin-based scale of *E. mulciber*. **b–f** Cu scales prepared via electroless deposition for **b** 5 min, **c** 10 min, **d** 15 min, **e** 20 min, and **f** 25 min, respectively. Scale bars: 1 μm. (Reproduced with permission [24] Copyright 2012, Wiley-VCH)

main ridges and rib structures collapse after the removal of original templates in H_3PO_4 (Fig. 5.7b). By contrast, when DT in-creased to 15 min, the surface of Cu butterfly scales coarsened as Cu grains grew larger with the increase in DT, generating an over-loaded metal layer on the wing surface. This phenomenon would be more obvious with a further increase in DT. When DT was prolonged to 25 min, the microstructures of Cu replicas were filled with Cu NPs and could hardly reserve their original morphologies. Figure 5.8 shows XRD results of Cu scales prepared with different DT. By reducing the scanning rate from 2°/min to 0.5°/min, we could accurately obtain (111) diffraction peak (see Fig. 5.8b). The average Cu grain size in response to different DT was 15.9, 16.9, 19.2, 22.4, and 28.6 nm, respectively (Fig. 5.8c). This tendency consists well with the SEM observations shown in Fig. 5.7. All these results prove that the thickness of metal layers can be effectively modified by changing the Cu DT. The morphology of Cu wings scales can thus be effectively controlled. By studying the SERS performance of Cu scales with deferring structures, coupling relationship between scale structures and SERS performance can be unveiled.

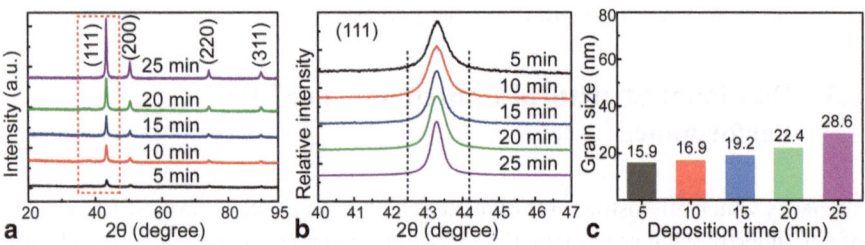

Fig. 5.8 a X-ray diffraction (XRD) results of Cu scales prepared with different deposition time (DT). **b** XRD results at (111) peak (scan rate: 0.5°/min). The peak intensities of different samples were normalized to unity for comparison. **c** Grain size calculated using the Scherrer formula. (Reproduced with permission [24] Copyright 2012, Wiley-VCH)

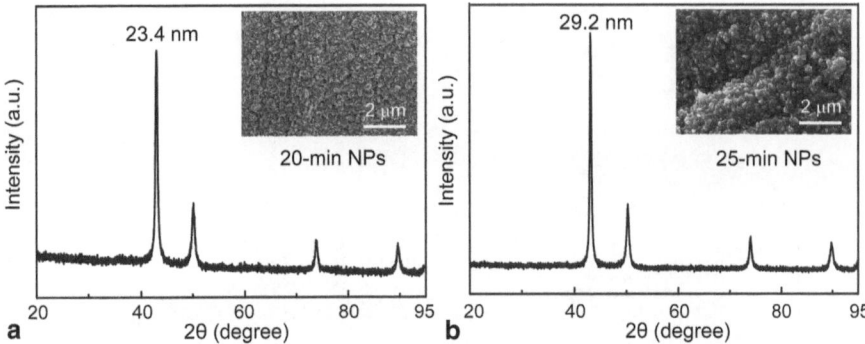

Fig. 5.9 Scanning electron microscope (SEM) and X-ray diffraction (XRD) analyses on Cu NPs as surface-enhanced Raman scattering (SERS) substrates. **a** and **b** are Cu NPs ground from Cu scales, which were synthesized by depositing Cu for 20 and 25 min, respectively. The calculated grain sizes are 23.4 and 29.2 nm for (**a**) and (**b**), respectively. (Reproduced with permission [24] Copyright 2012, Wiley-VCH)

A variety of Cu scale replicas with Cu DT of 10, 20, and 25 min were collected as an experimental group. Various films of Cu NPs ground from corresponding Cu replica were applied as a control group. Figure 5.9 shows SEM and XRD results of Cu NPs ground from Cu scales in liquid N_2. It confirms the structures of Cu scales were completely destroyed, while the size of Cu NPs hardly changed before (see Fig. 5.8c) and after the grinding process conducted in liquid N_2. Figure 5.10 shows the Raman spectra of a R6G solution (10^{-5} M) dripped and dried on these substrates. R6G Raman intensity at 1650 cm^{-1} was always stronger on Cu butterfly scales than on their as-ground NPs (Fig. 5.10b). Moreover, the R6G signals acquired on the over-coated Cu scale (DT 25 min, see Fig. 5.7f) decreased slightly after the grinding process. These results indicate that fine structures of butterfly scales dominate the SERS performance of metal scale replicas.

The structure of *E. mulciber* scales generally has three levels of periods. They can be described by the spacing between adjacent main ridges (P1), struts (P2), and ribs (P3). To compare the contribution of these periods on SERS performance, Cu butterfly scales shown in Fig. 5.7b–f were used as SERS substrates to detect R6G molecules (10^{-5} M). Figure 5.11g provides the Raman spectra of R6G molecules collected on these five types of Cu butterfly scales. As shown in Fig. 5.11h, the Raman intensity first increased with the Cu DT. When DT approached to 10 min, the strongest Raman signal was achieved on the corresponding sample (DT-10). The intensity was 3.5 times of magnitude that acquired on the specimen with a DT of 5 min (DT-5). As mentioned above, NPs on the surface of DT-5 were not continuous and could not retain the original structural morphologies as the Cu deposition was insufficient. When DT increased over 10 min, however, the Raman intensity started to decrease. Although the Raman intensity on DT-15 was only slightly lower than that on DT-10, the signals sharply decreased on DT-20 and DT-25. For the characteristic Raman peak at 1650 cm^{-1}, its intensity on DT-10 was 2.5 and 5 times stronger than that on sample DT-20 and DT-25, respectively. It should be noted that with the DT increase, structural features like ribs, struts, and main ridges were

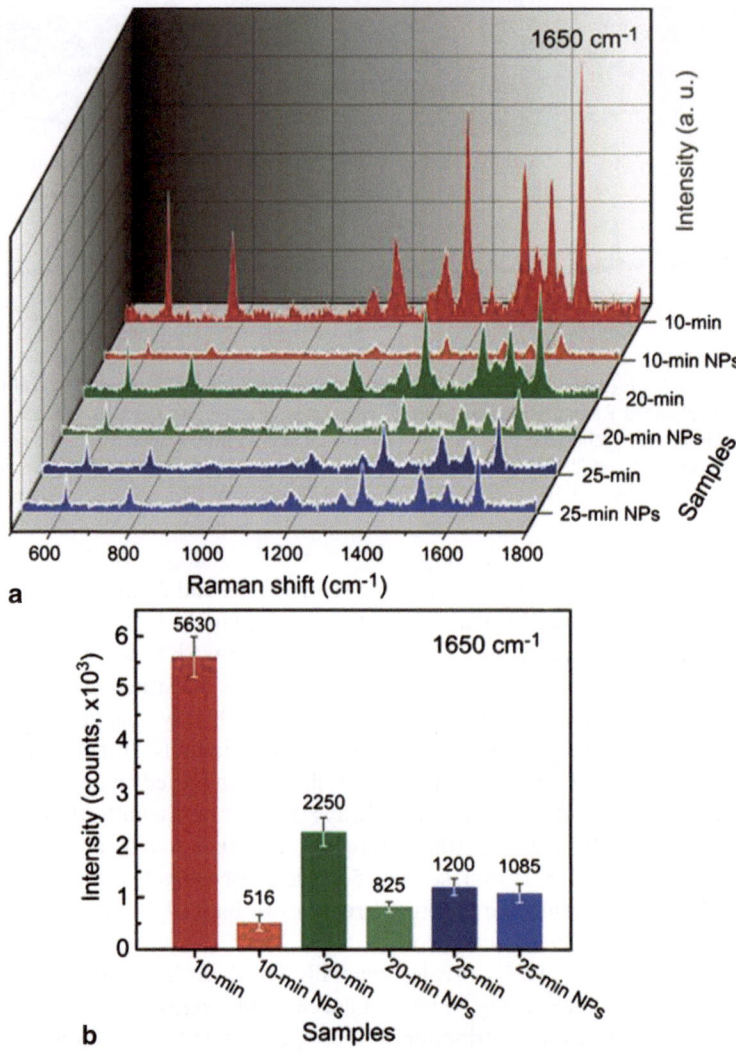

Fig. 5.10 **a** Surface-enhanced Raman scattering (SERS) spectra of R6G (10^{-5} M) acquired on various Cu replicas with or without the initial biomorphologies. **b** Raman intensities of R6G at 1650 cm^{-1}. (Reproduced with permission [24] Copyright 2012, Wiley-VCH)

successively destroyed. Although the grain size of Cu particles increased simultaneously, approaching to the ideal particle size of 50 nm for SERS [30], the NP-size-induced enhancement could not compensate the loss in the structure-induced amplification. Once again, these results confirmed that the significant SERS on Cu butterfly scales was mainly caused by 3D biostructures rather than the grain size or particle morphology of the metal [24, 31].

More detailed information can be deduced from Fig. 5.11:

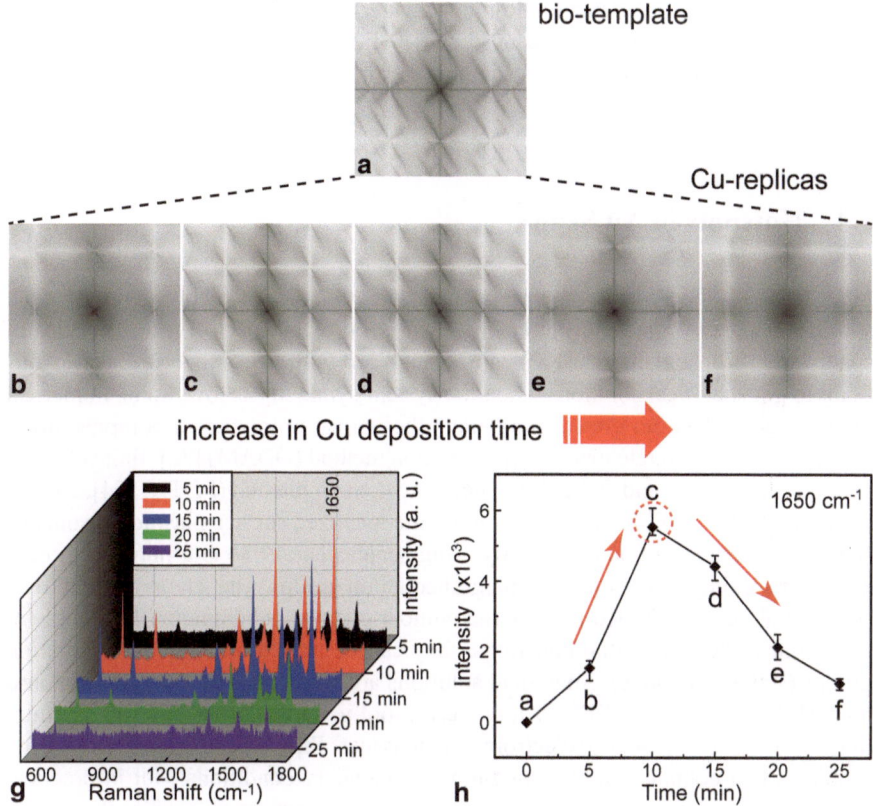

Fig. 5.11 a–f Scanning electron microscope (SEM) images treated using a fast Fourier-transform method. **a** Original scales. **b–f** Cu replicas with DT of **b** 5 min, **c** 10 min, **d** 15 min, **e** 20 min, and **f** 25 min, respectively. **g** Surface-enhanced Raman spectroscopy (SERS) spectra of 10^{-5} M R6G obtained on (**b–f**). **h** Raman intensities of (**g**) at 1650 cm^{-1}. (Reproduced with permission [24] Copyright 2012, Wiley-VCH)

- (1) When DT was 10 min (grain size 17 nm), the Raman intensity of R6G on Cu replicas with biostructures was 10 times that on the Cu NPs ground from these replicas. These NPs were structureless as their 3D morphologies had been mechanically destroyed (Fig. 5.9). This result suggests that the SERS effect caused by all the periods (P1 + P2 + P3) was 10 times of magnitude that caused by NPs themselves.

- (2) When DT reached to 20 min (grain size 22 nm), samples kept part of the periodicity (P1 and P2), with P3 lost. The signal intensity of R6G on DT-20 was thus 2.7 times of magnitude stronger than that on Cu NPs, suggesting that the SERS effect caused by P1 + P2 was 2.7 times stronger than that caused by NPs themselves.

Hence, it can be assumed that the Raman enhancement caused by P3 was four times stronger of magnitude than that caused by P1 + P2. It also means that for metal butterfly scales, the nanosized "rib" structure (P3) might play a dominant role in SERS performance [26–28].

5.4   Hotspots in Au Scales

In recent years, theoretical and experimental studies on the optical properties of metal structures achieved great development. Theoretically, the spectral characteristics of spherical or ellipsoidal metal NPs can be analyzed with Mie's theory. But it is difficult to get analytical solutions when the system symmetry is low. Some numerical methods were thus developed to simulate the optical properties of complex structures, such as the discrete dipole approximation method (DDAM) [32], finite element method (FEM) [33], and finite difference time domain method (FDTD) [34], etc.

In this section, FEM is applied to study the localized surface plasmon resonance (LSPR) generated on metal butterfly wing scales. The FEM is a numerical technique for solving differential and integral equation groups. The whole domain to be solved will be first divided into a large number of finite subdomains. To completely eliminate the differential equations, they can be turned into algebraic equation groups (stable situations). Analytical solutions are then obtained based on standard numerical methods. The FEM is a powerful tool to solve partial differential equations in complex region. For electromagnetic issues, FEM is suitable for solving the problems that contain complex mediums or boundary conditions. But in practical simulation processes, this method is limited by the performance of applied computers. Since the FEM ultimately turns differential problems into numerical linear algebra problems, this process will generate a large number of linear equations and the calculation scale depends on the as-divided subdomain numbers.

FEM has been realized via many commercial softwares. For example, COMSOL Multiphysics 3.5 is widely used in various fields of scientific calculations. This program can use FEM to efficiently conduct multiphysics simulations for physical problems described by partial differential equation groups. It helps researchers to obtain deep understandings in physical phenomena and can improve the credibility of experimental results as well. The RF module of COMSOL is used herein, which is a professional tool for the simulations on electromagnetic field related processes. This module has been specifically applied to simulate the scattering fields, scattering cross section of radars, electromagnetic propagations in Bloch–Floquet periodic array structures, negative refractions in metamaterials, transmission fields in 1D photonic structures, and the S-parameters and far-field properties of antennas, etc. [35, 36].

For the simulation process, some important parameters shall be set first. For example, optical characteristics of a medium are mainly described by its dielectric coefficients. Thus, appropriate dispersion model should be selected before simulation to accurately reflect the frequency dependence of the dielectric coefficient on radiation. For metals, the dielectric coefficient can be approximately described by Lorentz–Drude model [37]. The dielectric coefficients of some metals (e.g., Ag, Au,

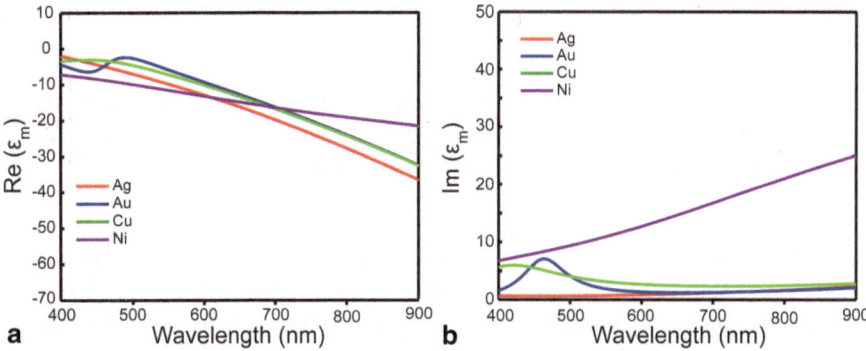

Fig. 5.12 Dielectric coefficients of Ag, Au, Cu, and Ni against different radiation wavelengths

Cu, and Ni) can thus be determined. Figure 5.12 shows the dielectric coefficients of Ag, Au, Cu, and Ni in response to different radiation wavelengths.

Once the Cu butterfly structures couple with incident lights, the spatial distributions of electromagnetic field will be rearranged. These processes were simulated using FEM. An incident plane wave was set to propagate perpendicularly to the scale surface, with a wavelength of 514.5 nm. The dielectric constant of air was one and the dielectric coefficient of Cu was obtained from Fig. 5.12. According to the structural parameters provided by SEM observations, periodically textured models of DT-10 and DT-20 were built to simulate the electromagnetic field distributions on Cu butterfly wing scales (Fig. 5.13). Detailed structural parameters are shown in Table 5.1.

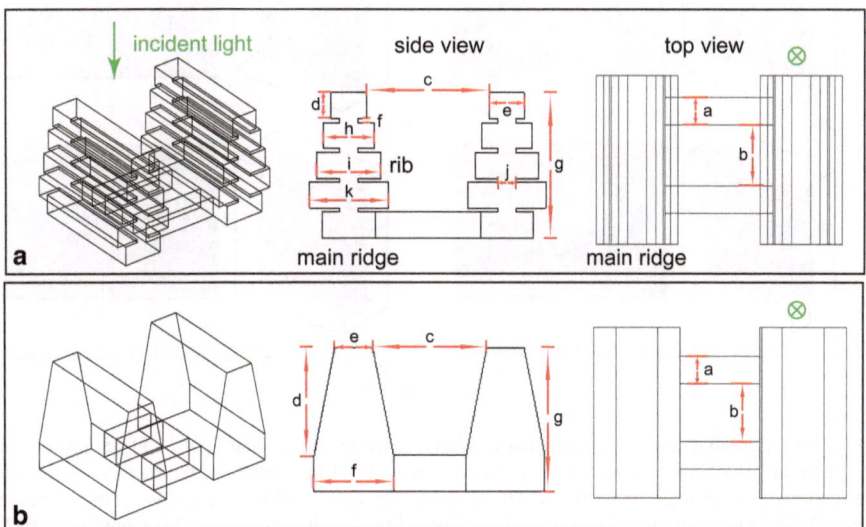

Fig. 5.13 Models for FEM simulations: **a** DT-10 with rib-structures and **b** DT-20 without rib-structures. Wavelength of the incident light is 514.5 nm according to the experimental setup. (Reproduced with permission [24] Copyright 2012, Wiley-VCH)

Table 5.1 Detailed dimensions applied in Fig. 5.13 for FEM simulations. (Reproduced with permission [24] Copyright 2012, Wiley-VCH)

	A (nm)	B (nm)	C (nm)	D (nm)	E (nm)	F (nm)	G (nm)	H (nm)	J (nm)	I (nm)	K (nm)
DT-10	200	300	680	144	200	20	800	280	360	100	440
DT-20	240	320	660	600	220	460	860	–	–	–	–

Figure 5.14a shows as-calculated electromagnetic field distributions on DT-10 and DT-20 with the excitation polarization perpendicular to the length direction of main ridges. Within rib structures, DT-10 strongly coupled with the incident light, producing high-density hotspots in the nanogaps of adjacent individual ribs. In comparison, there were no hotspots on DT-20 whose rib structures were mechanically destroyed. Similar results appeared (Fig. 5.14b) when the excitation polarization was parallel to the length direction of main ridges. It should be noted that the particle roughness was not considered in this simulation. The scattering by the rough surface could undermine the original polarization of incident lights. According to previous literatures [38–40], SERS enhancement is approximately proportional to the fourth power of the intensity of the local EM field. Thus, these simulation results indicate that the 3D Cu scales with fine rib structures can efficiently generate localized surface plasmons and substantially enhance EM fields in the rib-gaps. The enhanced EM fields provide additional hotspots especially arranged along the direction vertical to the scale surface, thus effectively enhancing the SERS performance.

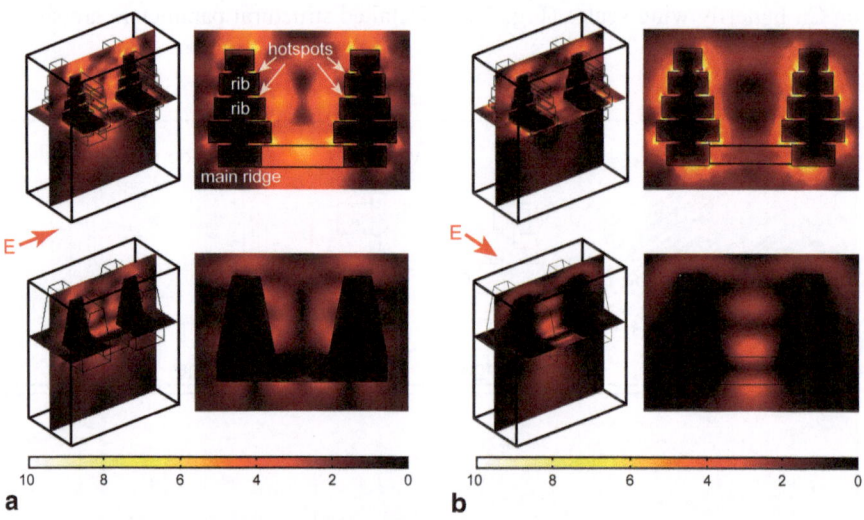

Fig. 5.14 Distribution of electromagnetic fields excited by an incident light (λ 514.5 nm) on Cu replicas of *E. mulciber*. The incidence polarizations are **a** perpendicular and **b** parallel to the length direction of main ridges, respectively. The scale bars denote the relative intensities of the scattered EM fields against the incident EM fields. (Reproduced with permission [24] Copyright 2012, Wiley-VCH)

5.5 Experimental Verification

This mechanism should be experimentally checked. Butterfly scales have a huge variety of morphologies, which allows us to select differently textured scales to prepare their Cu replicas. The applied scales included the ventral forewing scales of *E. mulciber* (referred to hereafter as B), dorsal hindwing scales of *E. mulciber* (C), ventral forewing scales of *Kallima inachus* (D), ventral forewing scales of *Thaumantis diores* (E)). The Cu DT was set to be 10 min, the optimized condition discussed above. Previous Cu scales replicated from *E. mulciber*'s dorsal forewing are referred to hereafter as A for comparison. Figure. 5.15 and 5.16show the SEM

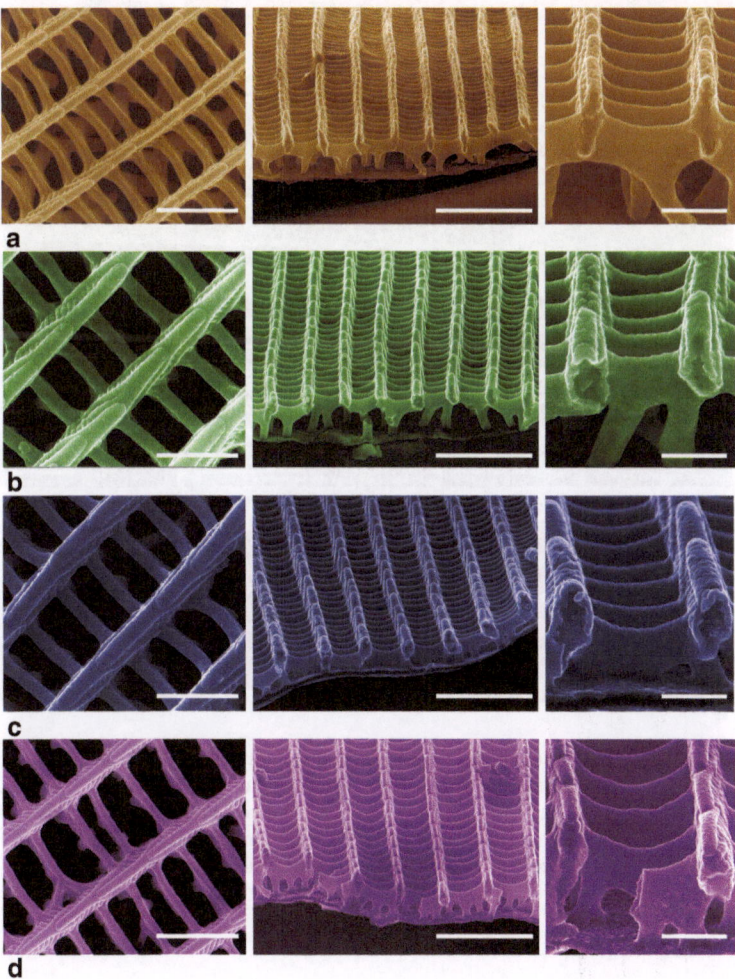

Fig. 5.15 Cu replicas with topologically similar morphologies inherited from natural scales on **a** ventral forewing surface of *E. mulciber*, **b** dorsal hindwing surface of *E. mulciber*, **c** dorsal forewing surface of *K. inachus*, and **d** dorsal forewing surface of *T. diores*. Scale bars: 2 μm for the left column, 5 μm for the middle column, and 1 μm for the right column. (Reproduced with permission [24] Copyright 2012, Wiley-VCH)

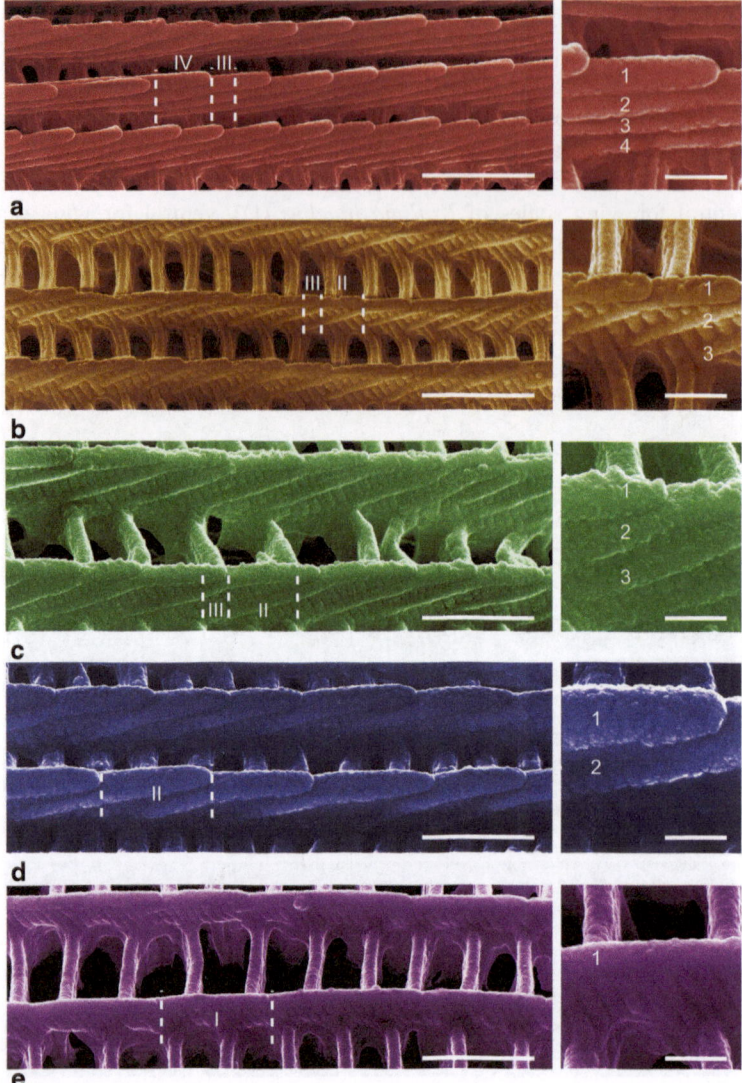

Fig. 5.16 Rib-structures in Cu replicas converted from natural scales on **a** dorsal forewing surface of *E. mulciber*, **b** ventral forewing surface of *E. mulciber*, **c** dorsal hindwing surface of *E. mulciber*, **d** dorsal forewing surface of *K. inachus*, and **e** dorsal forewing surface of *T. diores*. Stacking modes and numbers of ribs are marked herein. Scale bars: 2 μm for the left column and 500 nm for the right column. (Reproduced with permission [24] Copyright 2012, Wiley-VCH)

Table 5.2 Dimensions of the Cu scales in Fig. 5.15 and Fig. 5.16 [24]. Definition of these structures can be found in Fig. 5.1. Each datum was averaged from ten data measured at different spots

Scales	Window length (μm)	Window width (μm)	Ridge width (μm)
A	0.68	320	200
B	1.52	500	200
C	1.8	780	285
D	2.0	650	300
E	2.12	695	220

images of these Cu scales. The structural parameters of Cu wing scales are shown in Table 5.2. Although these scales are topologically similar (refer to the difference between the scales of *E. mulciber* and *Papilio paris*), the structural parameters of the ribs stacking on individual ridge are different. As shown in Fig. 5.16, the number of rib layers in A–E is 4/3, 3/2, 3/2, 2, and 1, respectively. Since the main ridges of A–E share the similar height, Scale A with the most rib layers has the smallest rib-gap, thus R6G on Scale A exhibited the strongest Raman signals (Fig. 5.17) [24, 41, 42]. Compared with Scale A, Scale B and C have fewer ribs stacking on their ridges, so the R6G signal on B and C was lower than that on A. In addition, although B and C have the same number of rib layers, the distance between the adjacent two ridges for B and C are 1.5 and 1.8 μm, respectively. In a certain illumination area, the density of hotspots produced by C is lower than that by B, so the R6G intensity on C was lower than that on B. It should be noted that the rib stacking number of Scale D is less than that of C and the spacing between the main ridges of D (2.0 μm) is larger than that of C (1.8 μm). Thus, the hotspot density on D was lower than that on C, giving rise to a lower R6G Raman intensity detected on D. Similarly, Scale E only has one rib layer, so it produced the weakest Raman signals. These results strongly verify the hypothesis proposed in the last section.

5.6 Summary

In this chapter, a variety of Cu wing scales with differing structures are prepared by adjusting the electroless DT. The influence exerted by different periodic structures on SERS properties is studied. Experimental results suggest that the scale ribs are the key contributor to SERS performance. This hypnosis is supported by the FEM simulation and have been experimentally proven on Cu scales with various rib structures. In conclusion, 3D Cu wing scales with fine rib structures efficiently generate localized surface plasmons and substantially enhance the electromagnetic fields in the rib gaps. This process provides hotspots especially arranged along the direction vertical to the scale surface and effectively increases the SERS performance. Moreover, a trace amount (10^{-8} M) of R6G and CV molecules is detectable on these low-cost Cu scales. Such sensitivity is even better than those of some as-reported Ag and Au substrates.

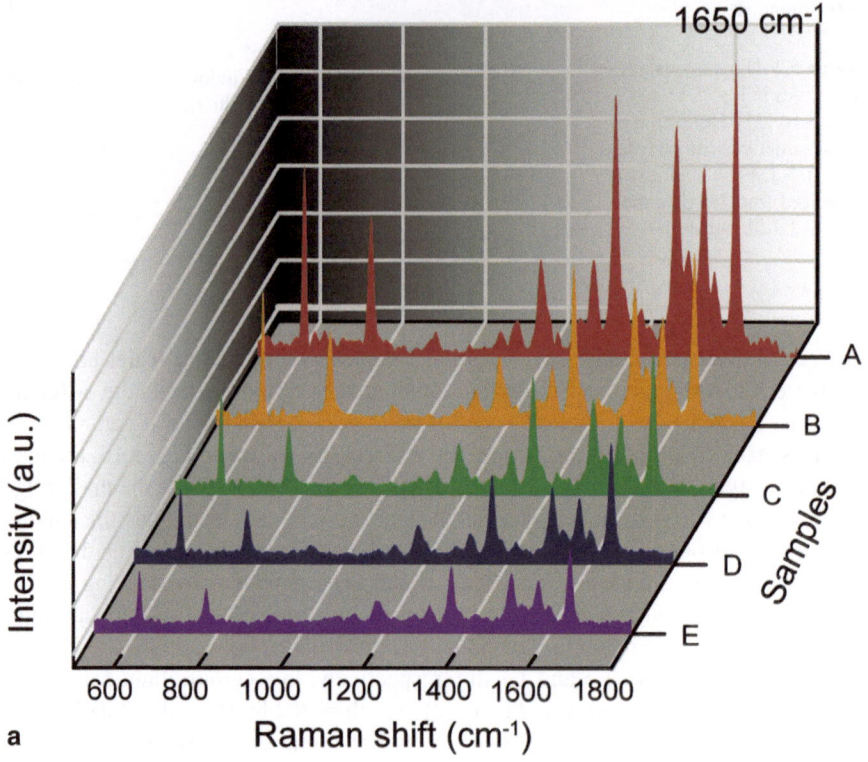

1650 cm^{-1}

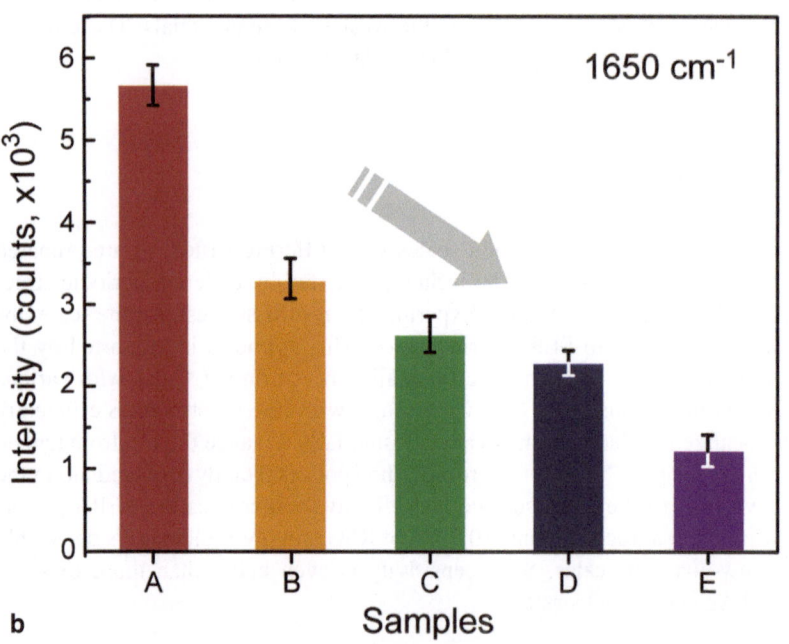

1650 cm^{-1}

Fig. 5.17 a Raman spectra of R6G molecules absorbed on Cu scales shown in Fig. 5.15 and 5.16. **b** Comparison of Raman intensities at 1650 cm^{-1}. (Reproduced with permission [24] Copyright 2012, Wiley-VCH)

References

1. Sharma B, Frontiera RR, Henry A-I et al (2012) SERS: Materials, applications, and the future. Mater Today 15:16–25
2. Otto A (1991) Surface-enhanced Raman scattering of adsorbates. J Raman Spectrosc 22:743–752
3. Moskovits M (2005) Surface-enhanced Raman spectroscopy: A brief retrospective. J Raman Spectrosc 36:485–496
4. Ding S-Y, Wu D-Y, Yang Z-L et al (2008) Some progresses in mechanistic studies on surface-enhanced Raman scattering. Chem J Chin Univ 29:2569–2581
5. Jensen L, Aikens CM, Schatz GC (2008) Electronic structure methods for studying surface-enhanced Raman scattering. Chem Soc Rev 37:1061–1073
6. Chang C, Clemente F, Kox R et al (2010) Raman scattered photon transmission through a single nanoslit. Appl Phy Lett 96:061108
7. Mclellan JM, Siekkinen A, Chen J et al (2006) Comparison of the surface-enhanced Raman scattering on sharp and truncated silver nanocubes. Chem Phys Lett 427:122–126
8. Duan H, Hu H, Kumar K et al (2011) Direct and reliable patterning of plasmonic nanostructures with sub-10-nm gaps. ACS Nano 5:7593–7600
9. He D, Hu B, Yao Q-F et al (2009) Large-scale synthesis of flexible free-standing SERS substrates with high sensitivity: Electrospun PVA nanofibers embedded with controlled alignment of silver nanoparticles. ACS Nano 3:3993–4002
10. Huang Z, Meng G, Huang Q et al (2010) Improved SERS performance from Au nanopillar arrays by abridging the pillar tip spacing by Ag sputtering. Adv Mater 22:4136–4139
11. Kang T, Yoo SM, Yoon I et al (2010) Patterned multiplex pathogen DNA detection by Au particle-on-wire SERS sensor. Nano Lett 10:1189–1193
12. Ko H, Singamaneni S, Tsukruk VV (2008) Nanostructured surfaces and assemblies as SERS media. Small 4:1576–1599
13. Abu Hatab NA Oran JM Sepaniak MJ (2008) Surface-enhanced Raman spectroscopy substrates created via electron beam lithography and nanotransfer printing. ACS Nano 2:377–385
14. Ou FS, Hu M, Naumov I et al (2011) Hot-spot engineering in polygonal nanofinger assemblies for surface enhanced Raman spectroscopy. Nano Lett 11:2538–2542
15. Jin ML, Pully V, Otto C et al (2010) High-density periodic arrays of self-aligned subwavelength nanopyramids for surface-enhanced Raman spectroscopy. J Phys Chem C 114:21953–21959
16. Crozier KB, Zhu WQ, Banaee MG et al (2011) Lithographically fabricated optical antennas with gaps well below 10 nm. Small 7:1761–1766
17. Wang W, Li Z, Gu B et al (2009) Ag@SiO$_2$ core-shell nanoparticles for probing spatial distribution of electromagnetic field enhancement via surface-enhanced Raman scattering. ACS Nano 3:3493–3496
18. Lu L, Sun G, Zhang H et al (2004) Fabrication of core-shell Au-Pt nanoparticle film and its potential application as catalysis and SERS substrate. J Mater Chem 14:1005–1009
19. Alvarez-Puebla RA, Agarwal A, Manna P et al (2011) Gold nanorods 3D-supercrystals as surface enhanced Raman scattering spectroscopy substrates for the rapid detection of scrambled prions. Proc Natl Acad Sci USA 108:8157–8161
20. Bryant GW, Garcia De Abajo FJ, Aizpurua J (2008) Mapping the plasmon resonances of metallic nanoantennas. Nano Lett 8:631–636
21. Wu WG, Qian C, Ni C et al (2011) Highly-ordered, 3D petal-like array for surface-enhanced Raman scattering. Small 7:1801–1806
22. Sweeney A, Jiggins C, Johnsen S (2003) Insect communication: Polarized light as a butterfly mating signal. Nature 423:31–32
23. Morehouse NI, Vukusic P, Rutowski R (2007) Pterin pigment granules are responsible for both broadband light scattering and wavelength selective absorption in the wing scales of pierid butterflies. Proc Roy Soc Lond B 274:359–366
24. Tan YW, Gu JJ, Xu LH et al (2012) High-density hotspots engineered by naturally piled-up subwavelength structures in three-dimensional copper butterfly wing scales for surface-enhanced Raman scattering detection. Adv Funct Mater 22:1578–1585

25. Vukusic P, Sambles JR, Lawrence CR et al (1999) Quantified interference and diffraction in single *Morpho* butterfly scales. Proc Roy Soc Lond B 266:1403–1411
26. Van Hooijdonk E, Barthou C, Vigneron JP et al (2011) Detailed experimental analysis of the structural fluorescence in the butterfly *Morpho sulkowskyi* (*Nymphalidae*). J Nanophoton 5:053525
27. Zamuner M, Talaga D, Deiss F et al (2009) Fabrication of a macroporous microwell array for surface-enhanced Raman scattering. Adv Funct Mater 19:3129–3135
28. Yu Q, Braswell S, Christin B et al (2010) Surface-enhanced Raman scattering on gold quasi-3D nanostructure and 2D nanohole arrays. Nanotechnology 21:355301
29. Shibu ES, Kimura K, Pradeep T (2009) Gold nanoparticle superlattices: Novel surface enhanced Raman scattering active substrates. Chem Mater 21:3773–3781
30. Mo Y, Mörke I, Wachter P (1984) The influence of surface roughness on the Raman scattering of pyridine on copper and silver surfaces. Solid State Comm 50:829–832
31. Tessier PM, Velev OD, Kalambur AT et al (2000) Assembly of gold nanostructured films templated by colloidal crystals and use in surface-enhanced Raman spectroscopy. J Am Chem Soc 122:9554–9555
32. Kelly KL, Coronado E, Zhao LL et al (2003) The optical properties of metal nanoparticles: The influence of size, shape, and dielectric environment. J Phys Chem B 107:668–677
33. Baker NA, Sept D, Joseph S et al (2001) Electrostatics of nanosystems: Application to microtubules and the ribosome. Proc Natl Acad Sci USA 98:10037–10041
34. Farjadpour A, Roundy D, Rodriguez A et al (2006) Improving accuracy by subpixel smoothing in the finite-difference time domain. Opt Lett 31:2972–2974
35. Khoury CG, Norton SJ, Vo-Dinh T (2009) Plasmonics of 3-D nanoshell dimers using multipole expansion and finite element method. ACS Nano 3:2776–2788
36. Oh J, Hart R, Capurro J et al (2009) Comprehensive analysis of particle motion under non-uniform ac electric fields in a microchannel. Lab on a Chip 9:62–78
37. Rodrigo SG, García-Vidal FJ, Martín-Moreno L (2008) Influence of material properties on extraordinary optical transmission through hole arrays. Phys Rev B 77:075401
38. Im H, Bantz KC, Lindquist NC et al (2010) Vertically oriented sub-10-nm plasmonic nanogap arrays. Nano Lett 10:2231–2236
39. Halas NJ, Lal S, Chang W-S et al (2011) Plasmons in strongly coupled metallic nanostructures. Chem Rev 111:3913–3961
40. Caldwell JD, Glembocki O, Bezares FJ et al (2011) Plasmonic nanopillar arrays for large-area, high-enhancement surface-enhanced Raman scattering sensors. ACS Nano 5:4046–4055
41. Deng XG, Braun GB, Liu S et al (2010) Single-order, subwavelength resonant nanograting as a uniformly hot substrate for surface-enhanced Raman spectroscopy. Nano Lett 10:1780–1786
42. Liberman V, Yilmaz C, Bloomstein TM et al (2010) A nanoparticle convective directed assembly process for the fabrication of periodic surface enhanced Raman spectroscopy substrates. Adv Mater 22:4298–4302

Chapter 6
Conclusions and Perspectives

The main results of this book are summarized as follows:

1) A group of metal materials with butterfly scale structures have been prepared via H_2 reduction, photoreduction, and electroless deposition. Among these approaches, electroless deposition conducted under mild temperature is proven to be the best for achieving a conformal replication of biostructures. Using this method, butterfly scales are successfully replicated at least in seven types of replicas (Ag, Au, Co, Cu, Ni, Pd, and Pt). As there are over 174,500 species of butterflies and moths, hundreds of thousands of metal submicrometer structures can be generated by this way. In addition, this method could also be extended to replicate many other biostructures (e.g., the exoskeleton of beetles) as chitin is one of the most abundant biomass on this planet.

2) Cu, Ag, and Au scale replicas can be used as surface-enhanced Raman scattering (SERS) substrates. These substrates exhibit excellent abilities in enhancing Raman signals. On Au scales, a trace amount of R6G down to 10^{-13} M in concentration can be detected. This value is one order of magnitude lower than that on commercial SERS substrates. Moreover, Au scales as SERS substrates show detection reproducibility comparable to their commercial counterparts, and can be prepared in common laboratories in 13 h. Considering that SERS substrates are consumables and Au scales are comparatively low in cost, they might bring high-performance SERS detections to ordinary laboratories.

3) The SERS performance originates especially from the periodic rib layers of butterfly scales. This structure can generate high-density hotspots along the third direction perpendicular to the scale surface, in turn significantly amplifying the Raman signals proportion to $|E|^4$. This finding not only helps to target the appropriate butterfly structure from countless candidates but also acts as proof-of-principles to the designing of high-performance SERS substrates.

To date, the rib structures introduced in this book have been employed in developing artificial SERS architectures. H. C. Jeon et al. generated hexagonally ordered

© Jiajun Gu, Di Zhang, and Yongwen Tan 2015
J. Gu et al., *Metallic Butterfly Wing Scales,* SpringerBriefs in Materials,
DOI 10.1007/978-3-319-12535-0_6

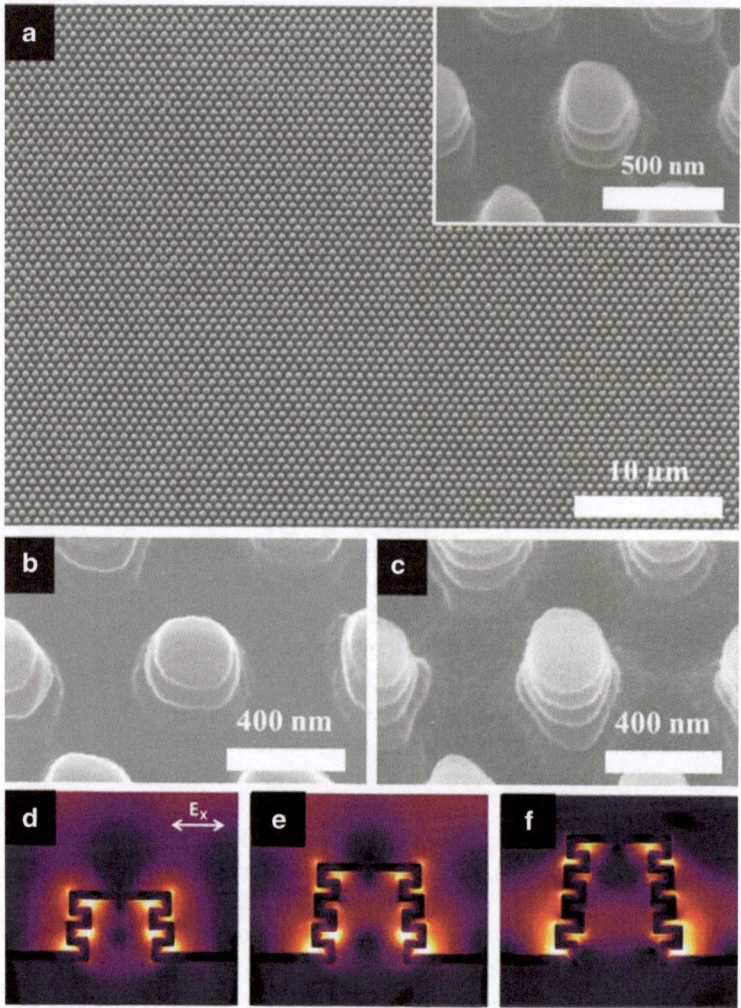

Fig. 6.1 a–c Scanning electron microscopy (SEM) images of ordered polymeric HORN arrays
with **b** two layers and **c** four layers. **d–f** Electromagnetic field intensity distribution in response to
the layer number. (Reproduced with permission [1] Copyright 2014, Wiley-VCH)

ridged nanostructure (HORN) arrays using dual interference lithography (Fig. 6.1;
[1]). The ridges on individual unit imitated the butterfly rib structures and could
provide hotspots along the direction vertical to the substrate surface [1]. Raman intensities of R6G significantly increased with the ridge numbers, which substantially
supports the mechanism provided in this book.

Additionally, because the superstructures scales can be produced using different
metals, they have potential for catalytic (Au, Pt, Pd), thermal (Ag, Au, Cu), electrical (Au, Cu, Ag), magnetic (Co, Ni), as well as optical (Au, Ag, Cu) applications.

Thus, a broad range of properties are yet to be explored on these metal replicas. Also, the developments of metal replicas of other biostructures are presently underway. We believe that more and more interesting results will be obtained from these studies.

Finally, there are many challenges to this new-born area too. One of the most difficult obstacles is that natural species will not develop their own structures according to human wills. To obtain perfect materials, the existing dimension of biostructures should be modified to be resonant with external factors, e.g., radiation wavelength. Although this goal can be ultimately achieved via gene-engineering, efforts to adjust biostructures through smart hydrogel coatings have been spared [2, 3], which is expected to generate more powerful functional materials in the near future.

References

1. Jeon HC, Jeon TY, Shim TS et al (2014) Direct fabrication of hexagonally ordered ridged nanoarchitectures via dual interference lithography for efficient sensing applications. Small 10:1490–1494
2. Zang XN, Ge YY, Gu JJ et al (2011) Tunable optical photonic devices made from moth wing scales: a way to enlarge natural functional structures' pool. J Mater Chem 21:13913–13919
3. Zang XN, Tan YW, Lv Z et al (2012) Moth wing scales as optical pH sensors. Sensor Actuat B Chem 166:824–828

Index